Human Apolipoprotein Mutants 2
From Gene Structure to Phenotypic Expression

NATO ASI Series

Advanced Science Institutes Series

A series presenting the results of activities sponsored by the NATO Science Committee, which aims at the dissemination of advanced scientific and technological knowledge, with a view to strengthening links between scientific communities.

The series is published by an international board of publishers in conjunction with the NATO Scientific Affairs Division

A	Life Sciences	Plenum Publishing Corporation
B	Physics	New York and London
C	Mathematical and Physical Sciences	Kluwer Academic Publishers
		Dordrecht, Boston, and London
D	Behavioral and Social Sciences	
E	Applied Sciences	
F	Computer and Systems Sciences	Springer-Verlag
G	Ecological Sciences	Berlin, Heidelberg, New York, London,
H	Cell Biology	Paris, and Tokyo

Recent Volumes in this Series

Series A: Life Sciences

Human Apolipoprotein Mutants 2

From Gene Structure to Phenotypic Expression

Edited by

C. R. Sirtori and G. Franceschini

Institute of Pharmacological Sciences
University of Milan
Milan, Italy

H. B. Brewer, Jr.

National Heart, Lung, and Blood Institute
National Institutes of Health
Bethesda, Maryland

and

G. Assmann

University of Münster
Münster, Federal Republic of Germany

Plenum Press
New York and London
Published in cooperation with NATO Scientific Affairs Division

Proceedings of a NATO Advanced Research Workshop on
Human Apolipoprotein Mutants:
From Gene Structure to Phenotypic Expression,
held March 27–30, 1988,
in Limone sul Garda, Italy

Library of Congress Cataloging in Publication Data

NATO Advanced Research Workshop on Human Apolipoprotein Mutants: from
Gene Structure to Phenotypic Expression (1988: Limone sul Garda, Italy)
    Human apolipoprotein mutants 2.
    (NATO ASI series. Series A, Life sciences; v. 167)
    "Proceedings of a NATO Advanced Research Workshop on Human Apolipopro-
tein Mutants: from Gene Structure to Phenotypic Expression, held March 27–30,
1988, in Limone sul Garda, Italy"—T.p. verso.
    "Published in cooperation with NATO Scientific Affairs Division."
    Includes bibliographies and index.
    1. Apolipoproteins—Congresses. 2. Gene expression—Congresses. 3. Human
chromosome abnormalities—Congresses. I. Sirtori, Cesare R. II. North Atlantic
Treaty Organization. Scientific Affairs Division. III. Title. IV. Series. [DNLM: 1.
Apolipoproteins—genetics—congresses. 2. Gene Expression Regulation—
congresses. 3. Mutation—congresses. QU 55 N28555h 1988]
QP99.3.A65N36   1988   573.2′292   89-8532
ISBN-13: 978-1-4615-9551-9   e-ISBN-13: 978-1-4615-9549-6
DOI: 10.1007/978-1-4615-9549-6

© 1989 Plenum Press, New York
Softcover reprint of the hardcover 1st edition 1989
A Division of Plenum Publishing Corporation
233 Spring Street, New York, N.Y. 10013

The pleasant community of Limone sul Garda provided outstanding
hospitality for a second NATO ARW dealing with apolipoprotein variants,
which are natures clues for the discovery of the physiological roles of
apolipoproteins in lipoprotein metabolism in normal subjects and
patients with specific dyslipoproteinemias.

Limone, the site of discovery of the first human apolipoprotein
mutant, apoA-I-Milano, provided a brilliant sunny spring venue for more
than 50 participants from both sides of the ocean.

The attendance at the colorful opening ceremony of the ARW was one
of the largest on record. Two members of the Italian government, the
Secretaries of Health and the Navy, gave the welcoming addresses. Six
television networks, two with national audiences, covered the
international workshop.

The Limone oracles provided a montage of insights gleamed from the
eyes of the clinican, the biochemist, and the molecular biologist. The
cumulative information on the molecular defects in lipoprotein
metabolism reviewed by this diverse group of investigators provided an
ever expanding horizon of new knowledge in this fast moving and some
times perplexing field. Clinical vignettes were presented on patients
from throughout the world including Canada (Connelly), Turkey (Schmitz),
and France (Infante) detailing the clinical sequelae of a defect in a
specific apolipoprotein. The clinical importance of Lp(a), a
lipoprotein relegated almost to obscurity for many years, has now taken

center stage. The increased risk of premature cardiovascular disease
associated with elevated plasma levels of Lp(a) will be a challenge to
the clinical investigator and physician since diet and the commonly used
drugs have little effects on plasma Lp(a) levels. Limone was also the
appropriate setting for the detailed description of the kinetic analysis
of apoA-I-Milano metabolism, and a final explanation for the variable
plasma levels of this mutant apolipoprotein (Roma, Gregg, Sirtori,
Franceschini).

During the workshop, the evolution of the lipoprotein transport
system was delineated in detail (Chapman), and an update on the quest
for the apoE or remnant receptor was presented (Beisiegel).

From the molecular biologist view point, the stop codon became a
focal point of interest. The introduction of the stop codon at the DNA
level provided the mechanism for several defects in apolipoprotein
biosynthesis (Fojo, Baggio). The herculean struggle of the lipoprotein
chemist vs the apoB protein was finally resolved, and nature begrudging
finally lifted the curtain on the mechanism for the biosynthesis of
apoB-100 and apoB-48 from the single apoB gene. Again the stop codon
provided the mechanism, this time being introduced at the RNA level by a
novel RNA editing mechanism. The elucidation of the covalent structure
of apoB and the discovering of the mechanism for the synthesis of
apoB-100 and apoB-48 are striking example of the power of the tools
available to the molecular biologist.

The superstars of gene polymorphism (Karathanasis, Ordovas,
Baralle) presented thoughtful overviews on the use as well as the
limitations of RFLP's in the analysis of genetic defects in single
kindreds and large scale population studies. The limitation of RFLPs as
a screening test for individual at risk for the development of premature
cardiovascular disease has now become apparent.

The venue, the hospitality of Limone, and the science at the second ARW was exemplary. I also have to personally express our appreciation to Cesare Sirtori and Guido Franceschini, our hosts, for the outstanding organization of the meeting, and their untiring efforts to collect the manuscripts for this book. We all look forward to the 3rd ARW on apolipoprotein mutants.

H. Bryan Brewer, Jr., M.D.

Chief, Molecular Disease Branch

National Heart, Lung, and Blood Institute

National Institutes of Health

Bethesda, Maryland, U.S.A.

CONTENTS

EVOLUTION AND CONTROL OF GENE EXPRESSION

HIGH DENSITY LIPOPROTEINS AND APO AI-MILANO

APOLIPOPROTEIN B

APOLIPOPROTEIN C-II AND C-III

APOLIPOPROTEIN GENES: ORGANIZATION, LINKAGE AND EVOLUTION

Sotirios K. Karathanasis

Laboratory of Molecular and Cellular Cardiology
Children's Hospital, and Department of Pediatrics
Harvard Medical School, Boston, MA 02115

INTRODUCTION

In the past three years all of the genes coding for the major human apolipoproteins have been cloned and sequenced. Using these cloned DNA sequences as probes for i) genomic blotting analysis of DNA from inter-species somatic cell hybrids, ii) in situ hybridization of metaphase chromosomes, iii) chromosomal "walking" by isolation and characterization of overlapping genomic clones, iv) genetic cosegregation of polymorphic DNA markers in family studies, and v) genomic blotting analysis of chromosomal DNA resolved by pulsed-field gel electrophoresis, it has been possible to determine the chromosomal localization and linkage relationships between all of these genes. Several conclusions have emerged. First, certain apolipoprotein genes are physically linked (i.e. clustered). For example, the apolipoprotein AI (apoAI), CIII (apoCIII) and AIV (apoAIV) genes are clustered within a 15 kilobase (kb) DNA fragment on the long arm of human chromosome 11 (1). Similarly, the apolipoprotein E (apoE), CI (apoCI) and CII (apoCII) genes are clustered within an approximately 50 kb DNA segment on the long arm of chromosome 19 (2). In contrast, the apolipoprotein AII (apoAII), B (apoB) and D (apoD) are dispersed on separate chromosomes, chromosomes 1, 2 and 3 respectively (reviewed in ref. 3).

Clustering of genes has major implications for their evolution, regulation of expression and patterns of inheritance at both the population and family levels (reviewed in refs. 3 and 4).

One of the focus points of our laboratory has been the study of the organization, linkage and evolutionary history of the apoAI, apoCIII and apoAIV genes. In this chapter I will summarize the concepts, experimental approaches and some of the results we have employed to show that these genes are closely linked and tandemly organized in the human genome. In addition, I will describe evidence indicating that this gene cluster was established more than 300 million years ago and that these genes are similarly organized in the genomes of all mammals, birds and possibly reptiles. The striking evolutionary conservation of this gene cluster implies that the tandem organization and spatial arrangement of these genes may play an important role in the regulation of their expression.

1. Protein sequence data suggest that apolipoproteins evolved from a common ancestor

Since 1977 it had been noticed that the amino acid sequences of apoAI, apoAII, apoCI and apoCIII, available at that time, exhibited statistically significant similarities (5). It had also been noticed that most of the apoAI amino acid sequence is made up by 22-residue repeats each of which is comprised of two 11-residue halves (5,6). Based on data indicating that an 11-residue consensus sequence derived from the amino acid repeats in apoAI is very similar to 11-residue segments in the apoAII, apoCI and apoCIII amino acid sequences, Barker and Dayhoff first proposed that all four of these proteins are derived from a common evolutionary ancestor (5). According to their model, a 33-base pair (bp) DNA segment in a primordial apolipoprotein gene coding for a small peptide, about 15-amino acid residues long, was internally duplicated to give rise to a new gene coding for a 26-amino acid residue long protein. This gene was subsequently duplicated to generate two independent genes. One of them was further duplicated and the resulting identical copies diverged by accumulation of nucleotide substitutions and by limited internal duplications to give rise to the current apoAII, apoCI and apoCIII genes. The other underwent multiple internal duplications to give rise to the current apoAI gene (see Fig. 3 in ref. 5).

Eleven years later and after complete nucleotide sequencing of all major human apolipoprotein genes, the basic concepts of this model are still valid and are currently used by several authors to elaborate on more refined versions of the model (4). Perhaps more important, however, is a prediction of this model, that is, that the coding region of the apoAI gene should contain multiple nucleotide repeats reflecting the internal duplication events occurred during its evolution.

2. The human apoAI gene contains several 66-nucleotide repeats

In 1983 we and others cloned and sequenced the human apoAI gene (7,8). Like most other eukaryotic genes the coding region of the apoAI gene is interrupted by intervening sequences (i.e. introns). Specifically the human apoAI gene contains three introns and, thus, four exons (7,8). More interestingly, however, an analysis of the nucleotide sequence of this gene by appropriate computer programs revealed that most of the sequences in exon 4 are made up by 66-bp repeats (see Fig. 3 in ref. 7). Although the actual number of these repeats depends upon the particular computer programs used, it is clear that each one of them codes for 22-amino acid long peptides with similar amphipathic helical characteristics. It should be noted that an important feature of amphipathic helices is the presence of hydrophobic and hydrophilic faces thought to be essential for the lipid-apolipoprotein interactions occurring on the surface of lipoprotein particles.

These results indicated that the apoAI gene evolved by intragenic duplications of a 66-bp DNA segment coding for a 22-residue long peptide with the potential of forming an amphipathic helical structure. Moreover, these results taken together with the proposed common evolutionary origin of apolipoproteins (see section 1 above) raised the possibility that the genes coding for these proteins are members of a multigene family generated by inter- and intra-genic duplication events of a fundamental DNA unit coding for a peptide with lipid binding potential.

3. Close physical linkage of the human apoAI and apoCIII genes

In general, gene duplications may occur in two different ways, that

is, by tandem duplication and by genome duplication. Tandem gene
duplication leads to the generation of clusters of identical genes.
Subsequent evolutionary divergence of these genes may lead to the rise of
genes coding for proteins with similar but distinct functions.

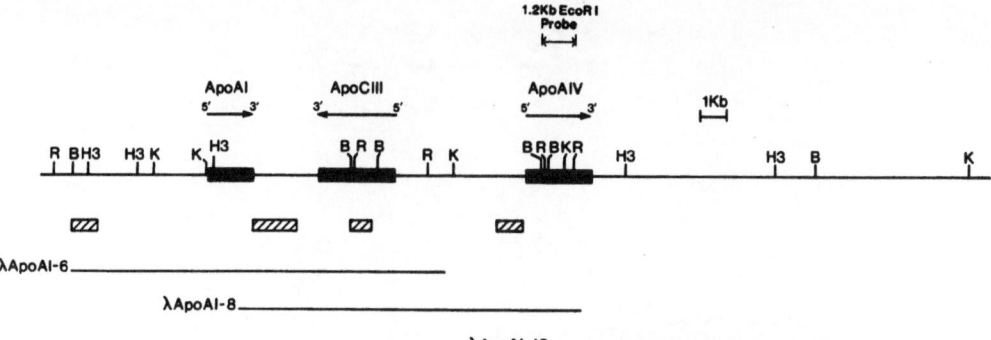

Fig. 1.   Restriction endonuclease map of a DNA segment containing the
           human apoAI, apoCIII, and apoAIV genes. The direction of
           transcription and the relative location of the genes are
           shown by arrows and solid boxes, respectively. Highly
           repetitive DNA sequences are indicated by shaded boxes. The
           extent and the relative location of genomic DNA fragments,
           contained in the specified λ clones, are shown by lines under
           the map. The nculeotide scale of the map is indicated by a
           solid bar. The relative location of a 1.2 kb EccRI fragment
           used for probe preparation is shown. Restriction
           endonuclease sites are indicated as follows: B, BamHI; H3,
           HindIII; K, KpnI; R, EcoRI.

     The possible common evolutionary origin of the apolipoprotein genes
and their likely generation by gene duplication events raised the
possibility that at least some of them may be physically linked. In
particular, the apoAI and apoCIII genes could be linked since an
inherited DNA rearrangement of the apoAI gene in certain patients with
premature atherosclerosis is associated with combined apoAI and apoCIII
deficiency in the plasma of these patients (9). To study this
possibility we isolated an apoCIII mRNA derived cDNA clone from an adult
human liver library (10). The DNA insert in this clone was subsequently
used as probe in hybridization blotting analysis of a genomic clone
(clone λApoAI-6, Fig. 1) containing the human apoAI gene. The resulting
autoradiogram (see Fig. 3 in ref. 10) showed that this genomic clone
contains, in addition to the apoAI gene, a DNA region with strong
homology to the apoCIII cDNA probe. These results together with data
obtained by DNA sequencing determinations of this DNA region revealed
that the apoCIII gene is located 2.6 kb to the 3' end of the apoAI gene
and that these two genes are transcribed in opposite directions (Fig. 1).

     These results strongly support the concept that the apolipoprotein
multigene family evolved by tandem gene duplication events and raise the
possibility that additional apolipoprotein genes may be physically linked
with the apoAI and apoCIII genes.

4. The human apoAI-CIII-AIV gene cluster

     To search the DNA regions flanking the apoAI and apoCIII genes for
the possible presence of additional apolipoprotein genes we used

3

chromosomal "walking" procedures to clone an approximately 30 kb DNA
segment containing both of these genes and approximately 5.5 kb of DNA
sequences flanking the 5' end of the apoAI gene and 16.5 kb of DNA
sequences flanking the 5' end of the apoCIII gene (1). The entire 30 kb

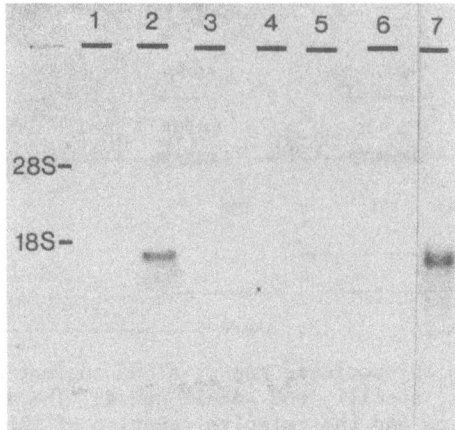

Fig. 2.   Total RNA (20 μg per lane) was isolated from human fetal
          liver (lane 1), intestine (lane 2), kidney (lane 3), heart
          (lane 4), brain (lane 5), muscle (lane 6), or adult liver
          (lane 7) and hybridized with a probe prepared using a 1.2 kb
          EcoRI fragment located 6.5 kb 5' to the apoCIII gene (Fig.
          1). The resulting autoradiogram is shown. The migration of
          28S and 18S ribosomal RNAs in indicated.

cloned genomic DNA segment is contained within the recombinant DNA
inserts of the overlapping genomic clones λApoAI-6, λApoAI-8, and λApoAI-
16, shown in Fig. 1. We reasoned that if other genes were present in the
flanking DNA regions of the apoAI and apoCIII genes then it should be
possible to detect mRNA transcripts of these genes by hybridization of
blotted RNA from various tissues with probes prepared from DNA fragments
derived from these regions. Several such probes were tested. The
autoradiogram obtained using a probe prepared with a 1.2 kb EcoRI
fragment located approximately 6.5 kb 5' to the apoCIII gene is shown in
Fig. 2. It shows that human fetal intestine and adult liver but not
fetal liver, kidney, heart, brain or muscle contain an approximately 1.8
kb mRNA transcript with extensive hybridization homology to this probe.
In subsequent work we used this probe to isolate several clones from an
adult human liver cDNA library (1,11). Comparison of the amino acid
sequence derived from the nucleotide sequence of one of these clones with
the amino acid sequences of other apolipoproteins revealed a highly
significant homology (61.8%) to the rat apoAIV (1,11). In addition,
analysis of this sequence for the presence of internal repeats revealed
that, similar to the human apoAI, most of this sequence is made up from
66 bp repeats coding for 22-amino acid repeats with amphipathic helical
characteristics (1,11).

     These results indicated that the apoAI, apoCIII and apoAIV genes are
clustered within the 15 kb DNA segment in the human genome and confirmed
the concept that the apolipoprotein multigene family evolved by tandem
gene duplication events.

5. Evolutionary history of the apoAI-CIII-AIV gene cluster.

The study of the organization and linkage relationships of the apoAI, apoCIII and apoAIV genes in the human genome (see above) provided strong support for the essential role of intra- and inter-genic duplication events in the evolution of the apolipoprotein multigene family. However, it also raised several questions with regard to the specific sequence of occurrence of these events. For example, protein and DNA sequence comparisons indicated that the apoAI and apoAIV genes are much more closely related to each other than either to the apoCIII gene (12). This is exemplified by the observation that both apoAI and apoAIV genes contain a much more extensive nucleotide sequence periodicity than the apoCIII gene. It would, therefore, appear that the apoCIII gene separated from the common evolutionary ancestor of all three of these genes before the intragenic duplication events that gave rise to the precursor of the apoAI and apoAIV genes. Thus, the apoAI and apoAIV genes were separated from each other after the apoCIII gene had been established. It is therefore difficult to understand why the apoCIII gene is located between the apoAI and apoAIV genes. It is also difficult to understand why the apoCIII gene is transcribed in the opposite direction than the apoAI and apoAIV genes (see Fig. 1). To explain this paradox Luo et al (12) proposed that after duplication of the apoAI gene to generate apoAIV, the apoCIII gene was inserted between apoAI and apoAIV by a DNA transposition event. However, there is no evidence of such an event. In addition, the same authors estimated by nucleotide sequence alignments and by determination of the rates of evolution of the apoAI and apoAIV genes that these two genes diverged from each other approximately 270 million years ago. Therefore, it would be expected that the genome of a species that branched off immediately before 270 million years ago should contain the apoAI and apoCIII genes transcribed in the same direction and it should not contain the apoAIV gene.

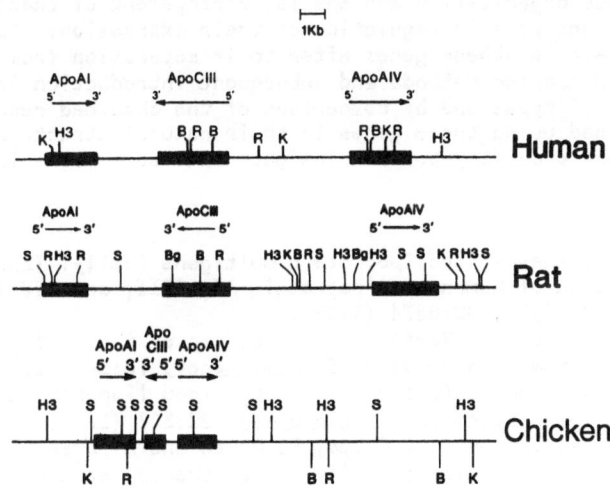

Fig. 3.   Organization of the apoAI-CIII-AIV gene cluster in the human, rat and chicken genomes. All symbols and notations are identical to that in Fig. 1 except that S indicates the presence of the restriction endonuclease site SacI.

We have recently addressed this issue by isolation and complete nucleotide sequence determination of all three apoAI, apoCIII and apoAIV genes in the rat (13) which branched off approximately 80 million years ago and the chicken (14) which branched off approximately 300 million years ago. The results are summarized in Fig. 3. They show that the relative location, size, direction of transcription and intron exon organization (not shown) of all three of these genes in the rat and in the chicken are remarkably similar to the corresponding genes in humans. These results indicate that the apoAI-CIII-AIV gene cluster was established more than 300 million years ago and that most likely all three of these genes are similarly organized in the genomes of all mammals and birds. Furthermore, it is currently believed that the mammalian line split from the reptile-bird line before reptiles and birds diverged from each other (15). Thus, since both mammals and birds shared the reptiles as a common ancestor and since the genome of both these two groups of animals contain the same organization of the apoAI, apoCIII and apoAIV genes it is very likely that these genes are also similarly organized in reptiles.

Clearly, more experimental data on the organization, linkage relationships, and nucleotide and derived amino acid sequences of all three of these genes in animals that branched off at even earlier evolutionary times will be needed to determine the evolutionary history of the specific events which occurred during generation of this gene cluster.

CONCLUSION AND FUTURE PROSPECTIVES

The processes controlling lipid metabolism are ultimately linked to the mechanisms that regulate the expression of genes coding for proteins involved in lipid transport. Regulation of expression of genes is intimately tied into the structural features of their surrounding DNA sequences. In particular, closely linked genes may influence each other's transcription by epigenetic mechanisms. Thus the remarkable evolutionary conservation of the apoAI-CIII-AIV gene cluster suggests that the tandem organization and spatial arrangement of these genes may play an important role in regulation of their expression. Future studies of the expression of these genes after their separation from each other by genetic engineering methods and subsequent introduction into appropriate cell types and by comparison of the obtained results with the results obtained using these genes in their natural structural arrangement may provide important insights into such mechanisms.

REFERENCES

1. S.K. Karathanasis, Apolipoprotein multigene family: Tandem organization of human apolipoprotein AI, CIII, and AIV genes, Proc. Natl. Acad. Sci., 82:6374 (1985).
2. O. Myklebost and S. Rogne, A physical map of the apolipoprotein gene cluster on human chromosome 19, Hum. Genet. 78:244 (1988).
3. A.J. Lusis, Genetic factors affecting blood lipoproteins: the candidate gene approach, J. Lipid Res. 29:397 (1988).
4. W.-H. Li, M. Tanimura, C-C Luo, S. Datta and L. Chan, The apolipoprotein multigene family: biosynthesis, structure-function relationships, and evolution, J. Lipid Res. 29:245 (1988).
5. W.C. Barker and M.O. Dayhoff, Evolution of lipoproteins deduced from protein sequence data, Comp. Biochem. Physiol. 57B:309 (1977).
6. A.D. McLachlan, Repeated helical pattern in apolipoprotein-A-I, Nature 267:465 (1977).
7. S.K. Karathanasis, V.I. Zannis and J.L. Breslow, Isolation and characterization of the human apolipoprotein A-I gene, Proc. Natl. Acad. Sci. USA 80:6147 (1983).
8. C.C. Shoulders, A.R. Kornblihtt, B.S. Munro and F.E. Baralle, Gene structure of human apolipoprotein AI, Nucleic Acids Res. 11:2827 (1983).

9. S.K. Karathanasis, E. Ferris and I.A. Haddad, DNA inversion within the apolipoproteins AI/CIII/AIV-encoding gene cluster of certain pateints with premature atherosclerosis, _Proc. Natl. Acad. Sci. USA_ 84:7198 (1987).

10. S.K. Karathanasis, J. McPherson, V.I. Zannis and J.L. Breslow, Linkage of human apolipoproteins A-I and C-III genes, _Nature_ 304:371 (1983).

11. S.K. Karathanasis, I. Yunis, and V.I. Zannis, Structure, evolution, and tissue-specific synthesis of human apolipoprotein AIV, _Biochem._ 25:3962 (1986). ·

12. C-C Luo, W-H Li, M.N. Moore and L. Chan, Structure and evolution of the apolipoprotein multigene family, _J. Mol. Biol._ 187:325 (1986).

13. I.A. Haddad, J.M. Ordovas, T. Fitzpatrick and S.K. Karathanasis, Linkage, evolution, and expression of the rat apolipoprotein A-I, C-III, and A-IV genes, _J. Biol. Chem._ 261:13268 (1986).

14. R. Sastry and S.K. Karathanasis, Linkage, evolution and expression of the chicken apolipoprotein AI, CIII, and AIV genes (in preparation).

15. R.E. Dickerson and I. Geis, "Hemoglobin", Benjamin/Cummings, Menlo Park, CA: (1983).

APOLIPOPROTEIN VARIATION: EFFECT ON PLASMA LIPID VARIABILITY

Gerd Utermann

Institut für Medizinische Biologie und Genetik
der Universität Innsbruck
Schöpfstrasse 41, A-6020 Innsbruck (Austria)

INTRODUCTION

Plasma lipid levels have a high variability within and in-between populations. Genetic factors are involved in controlling plasma lipid levels and explain more than 50 % of the variance of total plasma cholesterol levels in Caucasian populations. Single polymorphic genes affecting plasma lipid levels have only recently been identified (for review see Utermann, 1985, 1987a,b, 1988; Lusis, 1988).

Plasma lipids are transported and metabolized as lipoprotein complexes. The protein constituents of lipoproteins called apolipoproteins are important programmers of plasma lipid metabolism. Specific functions have been ascribed to certain members of this protein family. Apo A-I, apo A-IV and apo C-II are cofactors for lipolytic enzymes e.g. lecithin-cholesterol acyltransferase and lipoprotein lipase. Others are ligands for high affinity cell surface receptors and facilitate receptor mediated endocytosis of lipoproteins by cells e.g. apo B-100 and apo E (Brown and Goldstein, 1986). Allelic variation at apolipoprotein genes is likely to affect lipoprotein levels and phenotypes. Four of the known human plasma apolipoproteins are genetically polymorphic, namely apo A-IV, apo B, apo E, and the Lp(a) glycoprotein or apo(a)(Menzel et al., 1982; Utermann et al., 1977, 1982, 1987; Utermann, 1985, 1987a,b, 1988; Zannis and Breslow, 1981; Morganti et al., 1972; Young et al., 1986). The relationship of these apolipoprotein polymorphisms to plasma lipoprotein phenotypes and levels will be discussed here.

Polymorphism of apolipoprotein E

The gene for apo E on the long arm of chromosome 19 codes for a preprotein of 317 aminoacids that is converted cotranslationally to the mature 299 aminoacid plasma protein. In contrast to other apolipoprotein genes that are expressed in liver and intestine the apo E gene is not active in the intestine but in several other tissues throughtout the body and plays a key role in plasma lipid and possibly also in local lipid metabolism (Mahley, 1988). Apo E is a ligand with high affinity for the LDL-receptor and has been claimed to bind to the chylomicron-remnant receptor (Mahley, 1988). By virtue of its functional properties apo E is involved in three important pathways of lipoprotein metabolism: 1. the uptake of dietary cholesterol from chylomicrons by the liver, 2. the utilization of endogeneous cholesterol from VLDL-remnants and 3. the transport of cholesterol from peripheral cells to the liver (= reversed cholesterol transport). Human apo E is polymorphic (see Utermann, 1985, 1987b). There exist three common alleles designated $\varepsilon 2$, $\varepsilon 3$, and $\varepsilon 4$ that specify for isoforms E2, E3, and E4 in plasma. The resulting six phenotypes can be distinguished by electrofocussing of total delipidated serum or plasma followed by immunoblotting. The $\varepsilon 3$ allele is the most common in all populations studied so far, but there exist significant differences in apo E allele frequencies between populations. Amerindians lack the $\varepsilon 2$ allele and Black Sudanese have very high $\varepsilon 4$ allele frequencies (Table 1).

Genetic apo E isoforms differ by single aminoacid substitutions (Rall et al., 1984) that affect the functional properties of the protein. Apo E2 (158 arg → cys) has less than 2 % the binding activity to the LDL-receptor compared to the common apo E 3 form (Schneider et al., 1981). Apo E 4 (112 cys → arg) is indistinguishable in the in vitro binding from apo E 3 but exhibits an enhanced in vivo catabolism (Gregg et al., 1986). The functional differences between genetic apo E forms are reflected in strong, significant effects of the apo E gene locus on lipoprotein metabolism and lipid levels (Utermann et al., 1977, 1979; Davignon et al., 1988; Boerwinkle and Utermann, 1988). Since the discovery of the apo E polymorphism it is known that apo E 2/2 homozygotes exhibit a specific form of dyslipoproteinemia designated primary dysbetalipoproteinemia (Utermann et al., 1977). The metabolic abnormality is characterized by accumulation of chylomicron- and VLDL-remnants (ß-VLDL), but low LDL- and total cholesterol levels in plasma. The $\varepsilon 2$ allele does express itself not only in homozygotes but also in heterozygotes and moreover there is also an effect of the $\varepsilon 4$ allele on plasma lipid metabolism. In the German population the average effect of the $\varepsilon 2$ allele is to raize apo E levels by 9.5 mg/l and to lower apo B and total cholesterol by 9.5 mg/dl and 14 mg/dl respectively. The average effect of the $\varepsilon 4$ allele is opposite to that of $\varepsilon 2$ and roughly half in magnitude. The $\varepsilon 4$ allele decreases apo E levels by 1.9 mg/l and increases apo B and total cholesterol by 4.9 mg/dl and 7.1 mg/dl respectively (Boerwinkle and Utermann, 1988).

Principically the same observations were made in a variety of other ethnically and/or geographically distinct populations, e.g. Icelanders, Finns, Tyroleans and Japanese (see Utermann, 1988). However, studies in different ethnic groups from Singapore (Chinese, Indians, Malays) indicate that apo E allele effects may also depend on cultural background, e.g. nutrition (Utermann, 1987, 1988). In Indians and Malays the apo B and cholesterol raizing effect of the $\varepsilon 4$ allele is not present and the effect of the $\varepsilon 2$ allele is very week (E. Boerwinkle, N. Saha, G. Utermann, unpublished). The mechanism by which the allele $\varepsilon 2$ effects apo B, LDL and total cholesterol levels is only partially understood. We and others have proposed a mechanism that may explain the effect of apo E alleles on lipid metabolism (Utermann, 1985; Davignon et al., 1988; Boerwinkle and Utermann, 1988).

Some apo E 2/2 homozygotes develop a severe form of hyperlipoproteinemia called type III hyperlipidemia. Additional factors causing hyperlipidemia operate in such patients. The lipoprotein abnormality corresponds to the one described above as primary dysbetalipoproteinemia with the exception that affected subjects are grossly hyperlipidemic. The relationship of the apo E polymorphism with hyperlipidemia and in particular with type III hyperlipoproteinemia has been covered in several recent reviews (see Utermann, 1985, 1987b, 1988; Davignon et al., 1988).

Polymorphism of apolipoprotein A-IV

Apolipoprotein A-IV is synthesized by intestinal epithelial cells. It is a constituent of lymph chylomicrons and HDL but the major fraction of human apo A-IV is not associated with lipoproteins (Beisiegel and Utermann, 1979; Rosseneu, 1988). Apo A-IV activates the plasma enzyme lecithin-cholesterol acyltransferase in vitro (Steinmetz and Utermann, 1985) and the rat

Table 1. Relative Allele Frequencies of the Apo E Polymorphism
Among 16 Ethnically or Geographically Distinct Populations

| Populations | Alleles | | | |
	N	E2	E3	E4
Germany	1031	0.077	0.773	0.150
Tyrol	473	0.084	0.798	0.108
Netherlands	2000	0.082	0.751	0.168
France	223	0.130	0.742	0.128
Scotland	400	0.080	0.770	0.150
Iceland	176	0.063	0.773	0.165
Finland	615	0.040	0.773	0.227
Massachusetts	152	0.130	0.750	0.120
Canada	102	0.078	0.770	0.152
New Zealand	426	0.120	0.720	0.160
Japan	318	0.079	0.847	0.074
China	175	0.106	0.794	0.100
India	136	0.070	0.875	0.055
Malay	117	0.115	0.770	0.115
Amerindians	107	0.000	0.816	0.184
Sudanese	101	0.084	0.628	0.287

Data from Utermann, 1988; Boerwinkle et al., 1987; Klasen et al.,
1987; Sandholzer and Utermann, unpublished

protein has been shown to be a ligand for a specific liver cell receptor
(Dvorin et al., 1986). However, it's exact in vivo function remains unknown.
Human apo A-IV is genetically polymorphic (Utermann et al., 1982; Menzel
et al., 1982, 1988). Two common alleles A-IV^1 and A-IV^2 and some rare
alleles control the polymorphism. Apo A-IV^1 is the most frequent in all
populations studied so far (Table 2). We have recently developed a method
for typing apo A-IV by focussing of total delipidated plasma followed by
immunoblotting (Menzel et al., 1988). The effects of the apo A-IV alleles
on plasma cholesterol, triglyceride and HDL-cholesterol were investigated
in two populations, Tyroleans and Icelanders. In both populations we observed the same significant effect of the apo A-IV alleles on HDL-cholesterol levels. HDL-cholesterol was significantly increased in subjects with
the A-IV 2-1 phenotype. Genetic variation of the apo A-IV gene locus
accounts for 11 % of the total variability in HDL-cholesterol levels in
Tyroleans. Triglyceride levels were lower in subjects with the A-IV^2 allele.
The effect of the apo A-^{T}V polymorphism on triglycerides was significant
only in the Icelandic population. Lack of significance in the Tyrolean
sample probably is due to the fact that the Tyroleans blood donors were
none fasting. The effects cf the apo A-IV polymorphism observed by us are
consistent with our limited knowledge of the role of apo A-IV in chylomicron
and HDL metabolism.

Polymorphism of Lp(a) lipoprotein

The Lp(a) antigen was first described by Berg (1963) as an inherited
variant of beta-lipoproteins. Later it was shown that the Lp(a) antigenic
property resides in a distinct lipoprotein subpopulation called the Lp(a)
lipoprotein (Wiegandt et al., 1968) and that the concentration of this
lipoprotein varies extremely in between normal individuals (Albers and
Hazzard, 1974). Research on the Lp(a) lipoprotein was greatly stimulated

Table 2. Apo A-IV Polymorphism. Phenotype Frequencies and Effects on Lipid Levels

Population	N	Phenotype (%)			
		1-1	2-1	2-2	others
Tyrol	473	84.6	14.2	0.4	1.0
Germany	1000	85.1	14.0	0.5	0.4
Iceland	183	78.1	20.2	1.1	0.6
Ghana	99	89.9	0	0	10.1

Tyrol	Concentration (mg/dl)	
Phenotype	HDL-C	Triglyceride
1-1	53.0	229.0
2-1	56.9	204.7
2-2	31.4	165.5
Pooled	53.5	225.2
Prob.	0.003	0.34

Iceland		
Phenotype	HDL-C	Triglyceride
1-1	54.5	102.9
2-1	59.4	83.5
2-2	63.9	102.1
Pooled	55.7	98.6
Prob.	0.03	0.04

Data from Menzel et al., 1988, and Menzel, H.J., Sigurdsson, G., Boerwinkle, E., Utermann, G., submitted

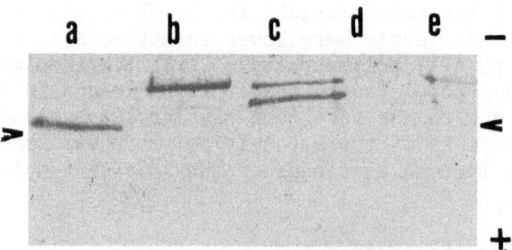

Fig. 1. SDS-PAGE of sera followed by immunoblotting with polyclonal anti-Lp(a) antibodies. The sera from five selected subjects represent the three principle patterns, a single band phenotype (a,b,e), double band phenotype (c) and null-type (d). Phenotypes are LpB (a), S2 (b,e), S1/S2 (c), null (d). Arrow indicates position of apo B-100 (M.W. ~ 550 KDa).

by reports that suggested a role of Lp(a) in the aetiology of atherosclerosis (see Rhoads et al., 1986). In Caucasians the distributions of Lp(a)
concentrations is highly skewed ranging from $<$ 1 mg/dl to $>$ 200 mg/dl and
with most subjects having low concentrations (mean $\approx$ 13 mg/100 ml).

The genetics of the quantitative Lp(a) trait remained largely unclear
and different genetic models accounting for the level differences have been
proposed (Sing et al., 1974; Hasstedt and Williams, 1986). The Lp(a) lipoprotein is assembled from two components, a particle corresponding to LDL
in protein (apo B-100) and lipid composition and from the high molecular
weight Lp(a) glycoprotein that is unique for the Lp(a) lipoprotein. The
Lp(a) glycoprotein has also been designated apo(a) even though it does not
have the properties of other apolipoproteins. LDL and apo(a) are linked by
a disulfide bridge in the complex.

We have recently developed a method to demonstrate the Lp(a) glycoprotein directly in plasma by SDS-gel electrophoresis followed by immunoblotting with poly- or monoclonal antibodies (Utermann et al., 1987, Kraft
et al., 1988, Fig. 1). With this technique at least six species of Lp(a)
glycoprotein differing in size from MW 400 KD-700 KD and designated LpF,
LpB, LpS1, LpS2, LpS3 and LpS4 can be distinguished (Table 3). A given individual exhibits either two, one, or none of these Lp(a) bands. Family
studies indicate that they are controlled by alleles at a single genetic
locus (Utermann et al., 1987,1988). Lp(a) types are significantly associated with Lp(a) concentration in plasma. Mean and median Lp(a) concentrations
are high in phenotypes with LpB, LpS1 and LpS2 isoforms. They are low in
phenotypes with LpS3 and S4 and very low or absent in the LpO type. Thus
there exists an inverse relationship between Lp(a) glycoprotein size and
Lp(a) lipoprotein plasma concentration. As much as 42 % of the variance in
Lp(a) plasma levels is explained by variation at the Lp(a) glycoprotein
locus (E.Boerwinkle and G. Utermann, unpublished). Recently apo (a) has
been sequenced both on the protein and on the cDNA level. This disclosed
and extremely high degree of homology of apo(a) with plasminogen (Eaton et
al., 1987, Mc Lean et al., 1987, Kratzin et al., 1987). Apo(a) contains
the protease domain, one copy of a kringle 5 structure and 37 repeats of
kringle 4. These findings have led to the speculation that genetic apo(a)
types differ by the number of kringle 4 structures.
The apo(a) gene locus, including its regulatory elements, probably is
the major gene locus determining Lp(a) plasma levels. The higher order
distribution of Lp(a) concentration in the population than results from
the allele specific effect on the level and the frequencies of the alleles
in the population (see Utermann, 1987a; Utermann et al., 1987).
There is growing evidence in our laboratory suggesting that Lp alleles
affect Lp(a) levels in an additive fashion. Heterozygotes on the average
have higher Lp(a) levels than the corresponding single band types. In a
heterozygous subject the two different Lp(a) species are on seperate Lp(a)
particles that may have different concentrations in plasma. The relationship between Lp(a) particle concentrations within a heterozygous subject
corresponds to the relation of apo(a) phenotype and Lp(a) level in the
population e.g. in a subject of phenotype S2/S4 there are more S2 than S4
containing Lp(a) particles. In a given phenotype there is however a wide
range of Lp(a) lipoprotein concentrations (Utermann et al., 1987, 1988)
and there is evidence for heterogeneity of levels within a phenotype (E.
Boerwinkle and G. Utermann, unpublished).

An association of Lp(a) lipoprotein concentrations with myocardial
infarction and stroke has been demonstrated in several studies. The mechanism by which Lp(a) lipoprotein exerts it's atherogenic effect is presently
unclear. In a recent editorial accompanying Mc Lean et al., (1987), Brown
and Goldstein (1987) have hypothezised that Lp(a) lipoprotein may provide
a link between the lipoprotein system and the clotting system. They suggest
that Lp(a) may adhere to fibrin following demage to the arterial wall and
may become entrapped as a component of the developing atherosclerotic
lesion. We have recently shown that apo(a) types not only affect Lp(a)

Table 3. Lp(a) Polymorphism: Isoforms, Allele Frequencies, and Effects on Lp(a) Levels

A

Isoform	M.W.(kDa)	Allele	Frequency
Lp(a) F	400	Lp^F	<0.005
Lp(a) B	460	Lp^B	0.007
Lp(a) S1	520	Lp^{S1}	0.027
Lp(a) S2	580	Lp^{S2}	0.123
Lp(a) S3	640	Lp^{S3}	0.125
Lp(a) S4	700	Lp^{S4}	0.163
–	–	Lp^O	0.553

B

Phenotype	N	Mean Lp(a) (mg/dl)
B	6	59 ± 21
S1	19	28 ± 20
S2	77	24 ± 22
S3	76	12 ± 9
S4	121	8 ± 7
0	138	3 ± 5

From Boerwinkle E., Menzel H.G., Kraft H.G., Utermann G., in preparation

levels but moreover have a significant effect on cholesterol levels. Variation at the apo(a) locus accounts for about 10 % of the variability in total cholesterol in the Tyrolean population (E.Boerwinkle, H.J.Menzel, H.G. Kraft, G. Utermann, submitted) and most of this is not mediated through Lp(a) levels. This finding may be an additional explanation for the association of Lp(a) lipoprotein with atherosclerosis.

CONCLUSIONS

There exist four genetically polymorphic systems in human, the apo A-IV, apo B (Ag), apo E and Lp(a) polymorphisms. Three of these have significant effects on plasma lipid metabolism. Apo E alleles contribute to the within population variance of total cholesterol-, LDL-cholesterol, remnant, apo B, and apo E-levels. Apo A-IV isoforms affect HDL-cholesterol and triglycerides. Apo(a) types have highly significant effects on Lp(a) lipoprotein and total cholesterol levels and explain about 40 % and 10 % of the total variance of these traits in the population. The profound effects of these gene loci on plasma lipoprotein metabolism implies that they are also genetic factors contributing to the risk for atherosclerosis. It will be a challanging goal of future research to unreveal the interaction of these and other loci in producing a lipoprotein phenotype.

REFERENCES

Albers, J. J., and Hazzard, W. R., 1974, Immunochemical quantification of human plasma Lp(a) lipoprotein, Lipids, 9:15

Beisiegel, U., and Utermann, G., 1979, An apolipoprotein homology of rat
        apolipoprotein A-IV in human plasma, Eur J Biochem., 93:601
Berg, K., 1963, A new serum type system in man: the Lp-system, Acta Pathol
        Microbiol Scand., 59:369.
Boerwinkle, E., Visvikis, S., Welsh, D., Steinmetz, J., Hanash, S. M., and
        Sing, C. F., 1987, The Use of Measured Genotype.Information in
        the Analysis of Quantitative Phenotypes in Man, Am J Med Genet.,
        27:567.
Boerwinkle, E., and Utermann, G., 1988, Simultaneous effects of the apo-
        lipoprotein E polymorphism on apolipoprotein E, apolipoprotein
        B and cholesterol metabolism, Am J Hum Genet., 42:104.
Brown, M. S., and Goldstein, J. L., 1986, A receptor mediated pathway for
        cholesterol homeostasis, Science, 232:34.
Brwon, M. S., and Goldstein, J. L., 1987, Plasma lipoproteins: Teaching
        old dogmas nwe tricks, Nature, 330:113.
Davignon, J., Gregg, R. E., and Sing, C. F., 1988, Apolipoprotein E poly-
        morphism and atherosclerosis, Arterioscl., 8:1.
Dvorin, E., Gordon, N. L., Benson, D. M., and Gotto Jr. A. M., 1986, Apo-
        lipoprotein A-IV. A determinant for binding and uptake of high
        density lipoprotein by rat hepatocytes, J Biol Chem., 261:15714.
Eaton, D. L., Fless, G. M., Kohr, W. J., Mc Lean, J. W., Xu, Q.-T., Miller,
        C.G., Lawn, R. W., and Scanu, A. M., 1987, Partial amino acid se-
        quence of apolipoprotein(a) shows that it is homologous to plas-
        minogen, Proc Natl Acad Sci. USA, 84:3224.
Gregg, R. E., Zech, L. A., Schaefer, E. J., Stark, D., Wilson, D., and
        Brewer, Jr. H. B., 1986, Abnormal in vivo metabolism of apolipo-
        protein E4 in humans, J Clin Invest., 78:815.
Hasstedt, S. J., and Williams, R. R., 1986, Three alleles for quantitative
        Lp(a), Genet Epidemiol., 3:53.
Klasen, E. C., Smit, M., De Kniff, P., Leuven, J. G., Kempen-Voogd, R.,
        and Havekes, L.,1987, Apolipoprotein E Phenotype and Gene Distri-
        bution in the Netherlands, Hum Hered., 37:340.
Kraft, H. G., Dieplinger, H., Hoye, E., and Utermann, G., 1988, Lp(a) pheno-
        typing by immunoblotting with polyclonal and monoclonal antibodies,
        Arterioscl.,.8:212.
Kratzin, H., Armstrong, V. W., Niehaus, M., Hilschman, N., and Seidel, D.,
        1987, Structural relationship of an apolipoprotein(a) phenotype
        (570 K Da) to plasminogen: Homologous kringle domains are linked by
        carbohydrate-rich region, Hoppe-Seyler's Z Biol Chem., 368:1533.
Lusis, A. J., 1988, Genetic factors affecting blood lipoproteins. The
        candidate gene approach, J Lipid Res., 29:397.
Mahley, R. W.,1988, Apolipoprotein E: Cholesterol transport protein with
        expanding role in cell biology, Science, 240:622.
Mc Lean, J. W., Tomlinson, J. E., Kuang, W., Eaton, D. L., Chen, E. Y.,
        Fless, G. M., Scanu, A. M., and Lawn, R. M., 1987, cDNA sequence
        of human apolipoprotein(a) is homologous to plasminogen, Nature,
        330:132.
Menzel, H. J., Kövary, P. M., and Assmann, G., 1982, Apolipoprotein A-IV
        Polymorphism in Man, Hum Genet., 62:349.
Menzel, H. J., Boerwinkle, E., Schrangl-Will, S., and Utermann, G., 1988,
        Human apolipoprotein A-IV polymorphism: frequency and effect on
        lipid and lipoprotein levels. Hum Genet., in press.
Morganti, G., and Beolchini, P. E., 1972, contribution to the genetics of
        serum ß-lipoprotein in man, Hum Genet., 16:307.
Rall, S. C. Jr., Weisgraber, K. H., and Mahley, R. W., 1982, Human apolipo-
        protein E. The complete amino acid sequence, J Biol Chem., 257:4171.
Rhoads, G. G., Dahlen, G., Berg, K., Morton, N. E., and Dannenberg, A. L.,
        1986, Lp(a) lipoprotein as a risk factor for myocardial infarction,
        JAMA, 256:2540.
Rosseneu, M., Michiels, G., de Keersgleter, W., Bury, J., De Slypere, J. P.,
        Dieplinger, H., and Utermann, G., 1988, Quantification of human

apolipoprotein A-IV by "Sandwich"-type enzyme linked immunosorbent assay, Clin Chem., 34:739.

Schneider, W. J., Kovanen, P. T., Brown, M. S., Goldstein, J. L., Utermann, G., Weber, W., Havel, R. J., Kotite, L., Kane, J. P., Innerarity, T. L., and Mahley, R. W., 1981, Familial dysbetalipoproteinemia. Abnormal binding of mutant apoprotein E to low density lipoprotein receptors of human fibroblasts and membranes from liver and adrenals of rats, rabbits, and cows, J Clin Invest., 68:1075.

Sing, C. F., Schultz, J. S., and Shreffler, D. C., 1974, The genetics of the Lp-antigen II, Ann Hum Genet., 38:47.

Steinmetz, A., and Utermann, G., 1985, Activation of lecithin:cholesterol acyltransferase by human apolipoprotein A-IV, J Biol Chem., 260: 2258.

Utermann, G., Hees, M., and Steinmetz, A., 1977, Polymorphism of apolipoprotein E and occurance of dysbetalipoproteinemia in man, Nature 604.

Utermann, G., Pruin, N., and Steinmetz, A., 1979, Polymorphism of apolipoprotein E. III: Effect of a single polymorphic gene locus on plasma lipid levels in man, Clin Genet., 15:63.

Utermann, G., Feussner, G., Franceschini, G., and Steinmetz, A., 1982, Genetic Variants of Group A Apolipoproteins. J Biol Chem., 257: 501.

Utermann, G., 1985, Genetic polymorphism of apolipoprotein E - impact on plasma lipoprotein metabolism, in: "Diabetes, obesity and hyperlipidemias-III," G. Crepaldi, ed., Elsevier Science Publishers, New York, pp 1.

Utermann, G., Menzel, H. J., Kraft, H. G., Duba, H. C., Kemmler, H. G., and Seiß, C. 1987, Lp(a) glycoprotein phenotypes. Inheritance and relation to Lp(a)-lipoprotein concentration in plasma, J Clin Invest., 80:458.

Utermann, G., 1987a, Apolipoproteins, quantitative lipoprotein traits and multifactorial hyperlipidemia, in:"CIBA Foundation Symposium 130", Wiley, Chichester, pp 52.

Utermann, G., 1987b, Apolipoprotein E polymorphism in health and disease, Am Heart J., 113:433.

Utermann, G., Duba, C., and Menzel, H. J., 1988, Genetics of the quantitative Lp(a) lipoprotein trait, Hum Genet., 78:47.

Utermann, G., 1988, Apolipoprotein variation: Effect on plasma lipid variability, J Inher Metab Dis., in press

Wiegandt, H., Lipp, K., and Wendt, G., 1968, Identifizierung eines Lipoproteins mit Antigenwirksamkeit im Lp-System, Hoppe-Seyler's Z Physiol Chem., 349:489.

Young, S. G., Bertics, S. J., Curtiss, L. K., Casal, D. C., and Witztum, L., 1986, Monoclonal antibody MB 19 detects genetic polymorphism in human apolipoprotein B, Proc Natl Acad Sci. USA, 83:1101.

Zannis, V. I., and Breslow, J. L., 1981, Human very low density lipoprotein apolipoprotein E isoprotein polymorphism is explained by genetic variation and post-translational modification, Biochemistry, 20:1033.

EXPRESSION OF HUMAN APO AI, AII AND CII GENES IN

PRO- AND EUCARYOTIC CELLS

W. Stoffel, E Binczek, A. Haase, and C. Holtfreter

Institut für Physiologische Chemie der Universität zu Köln
Joseph-Stelzmann-Str.52
D-5000 Köln 41

The recent short lived cloning boom of human serum apolipoprotein which finally made available cDNA clones and genomic clones of the ten serum lipoproteins has led to the confirmation of known polypeptide sequences as for AI, AII, CI-CIII and the elucidation of unknown primary proteins sequences, the most remarkable apo B and apo A. On the basis of these cloning experiments eventually the functionally and physiologically more important questions may be answered by expression studies of wild type apolipoprotein cDNA or genomic clones and particularly of mutagenized apolipoprotein genes.

I would like to report on studies which give three selected examples of the enormous potential of molecular biological methods in support of a new area in serum lipoprotein research.

1) The first topic is concerned with *in vitro* transcription-translation studies with wild type and mutagenized human apo AI in which we studied the cotranslational translocation of the preproform of apo AI cDNA wild type and of two mutants in which by site-directed mutagenesis with the gapped duplex method

a) the peculiar $Gln^{-2}-Gln^{-1}$ C-terminus of the prosequence and the $Gln^{-8}-Ala^{-7}$ C-terminus of the presequence was transposed and

b) the hexapeptide prosequence was deleted.

Among the primary translation products of the serum apolipoprotein RNAs only apo AI and AII carry a hexa- and pentapeptide proform respectively interposed between signal peptide and mature form. Nothing is known about the function of the propeptide.

Figure 1 schematically represents the expression vector for the wild type and mutant apo AI cDNA, the apo AI insert with 13 bp at the 5' end, 801 bp coding region and 75 at the 3' end, and figure 2 the mutations described before.

The synthetic nucleotides used in the gapped duplex mutagenesis are indicated below the mutant preprosequences. The mutations and the correct orientation of the inserts were confirmed by restriction and DNA sequence analysis, figures 3 a, b and 4 a, b. Transcription-translation *in vitro* of the wild type and of mutant 1 clones, combined with the translocation and processing endoplasmic reticulum membranes led to the results summarized in the autoradiograms of figure 5.

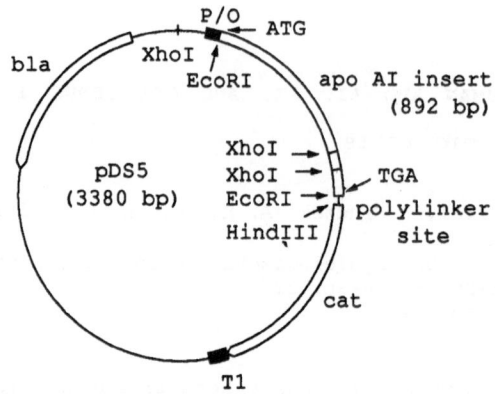

Fig. 1. Apo AI/pDS5 clone.

Normal sequence (**wild type**)

                presequence   prosequence

DNA	GGG AGC CAG **GCT**	CGG CAT TTC TGG CAG **CAA**	GAT GAA CCC
	-10 -7	-6 -1	+1
protein	Gly Ser Gln **Ala**	Arg His Phe Trp Gln **Gln**	Asp Gln Pro

Mutated sequence (**mutant 1**)

oligonucleotide	GGG AGC CAG **CAG**	CGG CAT TTC TGG CAG **GCT**	GAT GAA CCC
	-10 -7	-6 -1	+1
protein	Gly Ser Gln **Gln**	Arg His Phe Trp Gln **Ala**	Asp Gln Pro

Mutated sequence (**mutant 2**)

DNA	GGG AGC CAG **GCT**	-----------------------	GAT GAA CCC
	-10 -7	-6 -1	+1
protein	Gly Ser Gln **Ala**	missing prosequence	Asp Gln Pro

Fig. 2. Mutagenized sequences of apo AI

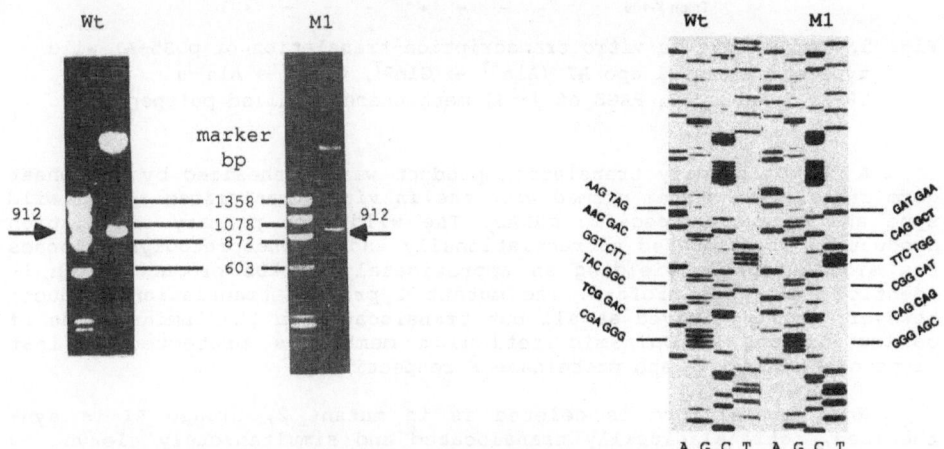

Fig. 3. a) *Hind*III/*Xho*I restriction analysis of
pDS5-AI wild type (Wt) and mutant 1 (M1).
b) Nucleotide sequence of wild type and mutant 1 of apo AI.

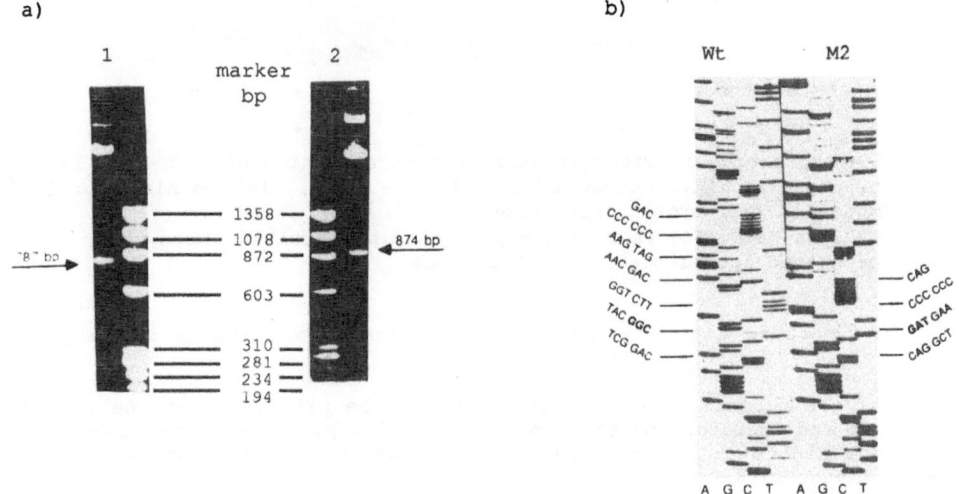

Fig. 4. a) Restriction analysis of pDS12-AI mutant 2 with
deleted prosequence of apo AI wild type.
1 *Bam*HI/*Xho*I restriction
2 *Eco*RI restriction
b) Nucleotide sequence of apo AI wild type (Wt) and
apo AI mutant 2 (M2) around deleted prosequence.

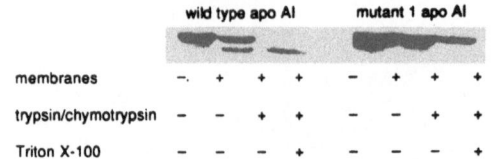

wild type apo AI   mutant 1 apo AI

membranes	−.	+	+	+	−	+	+	+
trypsin/chymotrypsin	−	−	+	+	−	−	+	+
Triton X-100	−	−	−	+	−	−	−	+

Fig. 5. Comparative *in vitro* transcription-translation of pDS5-AI wild
type and mutant 1 apo AI (Ala^{-7} → Gln^{-7}, Gln^{-1} → Ala^{-1}).
10-15 % NadDodSO$_4$ PAGE of [^{35}S] methionine labelled polypeptides.

A 31 kDa primary translation product was synthesized by the wheat
germ translation system primed with the *in vitro* transcribed apo AI wild
type and mutant 1 specific mRNAs. The wild type primary translation
product is translocated cotranslationally and by endoproteolysis looses
the signal peptide yielding an approximately 29 kDa product which is
identical with the proform. The mutant 1 primary translation product,
however, is not cleaved at all but translocated in the luminal side of
canine or dog endoplasmic reticulum membranes protected against
trypsin/chymotrypsin and proteinase K respectively.

When the proform is deleted as in mutant 2, preapo AI is synthesized, cotranslationally translocated and simultaneously cleaved by
signal-peptidase to mature apo AI with MW 28.5 kDa, figure 6. The
immunoprecipitated methionine- and proline-labelled apo AI forms were
radiosequenced by Edman degradation over 30 cycles and thereby the
structures confirmed.

Wt   M1   M2   Wt   M1   M2   Wt   M1   M2

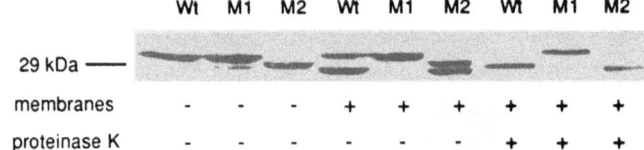

29 kDa —									
membranes	−	−	−	+	+	+	+	+	+
proteinase K	−	−	−	−	−	−	+	+	+

Fig. 6. Comparative *in vitro* transcription-translation of pDS5-AI wild
type apo AI (Wt), mutant 1 (M1, Ala^{-7} → Gln^{-7}, Gln^{-1} → Ala^{-1}) and
mutant 2 (M2, prosequence deleted).
10-15 % NadDodSO$_4$ PAGE of [^{35}S] methionine labelled, immunoprecipitated apo AI-specific polypeptides.

From these two site-directed mutagenesis experiments we propose
that

a) the proform is of no relevance for the processing of the signal
sequence and therefore neither contributes to a putative linear topogenic
sequence nor to a topogenic site in a folded tertiary structure around
the cleavage site,

b) the sequence following the signal peptide either the pro- or N-
terminal mature sequence, both very different in the case of apo AI, are
without influence on the processing.

c) information is inherent in the signal sequence itself and the
correct C-terminus.

The function of the prosequence of apo AI may be different from
that of apo AII. In the latter case the deletion leads to a cleavage by
the signal peptidase within the N-terminus of the mature sequence at the
C-terminus of Ala^{+2} instead of Gly^{-1} as shown by J. Gordon *et al.* Here

however the accurate cleavage behind the C-terminal Ala^{-1} of the pre-sequence of apo AI is preserved.

We search for functions different from that of a simple spacer between signal and mature sequence. The prosequence may intracellularly be of importance for the dichotomy in the posttranslational pathway to a secretory protein or intra- and extracellularly it may be required for the targeting process required for the assembly with lipids to the supramolecular structures of the still undefined secretory primary lipoprotein particle.

These mutants, equipped with the regulatory sequences from the genomic clone described later, can now be used for injection experiments of oocytes and to establish transfected cell lines with transient or per-manent expression of the apo AI and AII gene, cell lines which allow to address these and other questions.

2) In the second part I want to describe results of *in vivo* expres-sion experiments with genomic apo AI and AII DNA and the expression and secretion products analyzed.

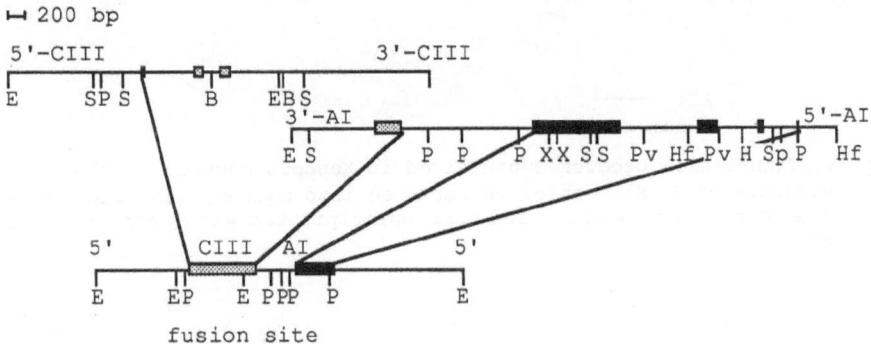

Fig. 7. Construction of fused apo CIII and apo AI genomic DNA with its restriction sites.

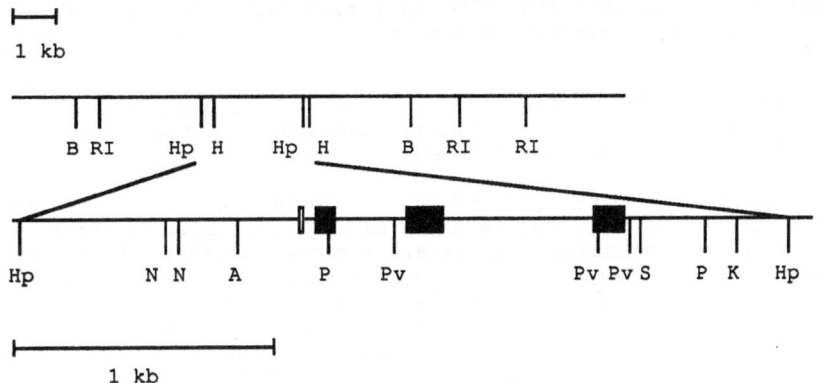

Fig. 8. Apo AII gene organisation with restriction sites.

A fused genomic apo AI-CIII clone of approximately 10 kb in λgt was constructed (figure 7) and an apo AII clone isolated (figure 8). Details of the characterization of both clones and mutants of the apo AII genomic clone for the study of regulatory regions are not shown.

OOCYTE INJECTION EXPERIMENTS

λAI and AII genomic DNA was injected into the nuclei of *Xenopus laevis* oocytes and the translation products synthesized in the presence of [^{35}S] methionine analyzed (figure 9). In addition ^{14}C acetate was given as a precursor for labelling lipids of the oocyte, newly synthesized in the incubation period and complexed with the secreted apolipoprotein.

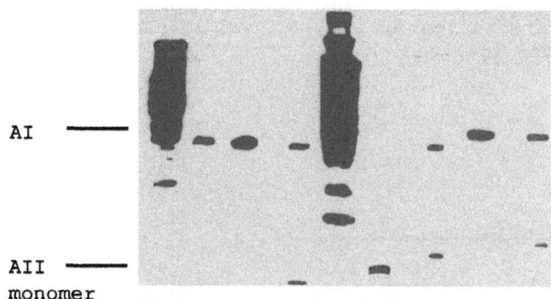

Fig. 9. Translation product synthesized in Xenopus oocytes in the presence of [^{35}S] methionine secreted into medium, separated by cell gradient centrifugation and immunoprecipitated with anti apo AI and AII.

The medium of oocyte injected with apo AI-CIII DNA was analyzed by CsCl-gradient centrifugation and the fractions analyzed for density [^{35}S] methionine labelled apo AI by immunoprecipitation and gradient (10-15 %) PAGE and autoradiography, and by radio thin layer chromatography for neutral and complex lipids.

These results were further confirmed by radio sequencing (30 Edman cycles) of [^{3}H] proline- and [^{35}S] cysteine-labelled translation products of AI and AII RNA respectively that the proforms of apo AI and AII are secreted.

Simultaneous injection of apo AI and AII genomic DNA into oocyte nuclei led to the secretion of the two proforms, but they were separated in two densities in the CsCl gradients.

Surprizing results appeared in the lipid analysis. The labelling pattern of the oocyte lipids is shown as TLC radio scan in figure 10. No labelled newly synthesized lipids are secreted (figure 10 A). However, oocytes expressing the apo AI gene secreted apo AI selectively associated with newly synthesized phosphatidylethanolamine (80 %) and neutral lipid (20 %), figure 10 B. Apo AII gene expression led to the secretion of a particle which banded at density 1.10-1.21 g/ml with proapo AII complexed with phosphatidylethanolamine and lyso-PE and neutral lipids (figure 10 C, D).

The simultaneous expression yielded particles of density 1.19-1.21 g/ml with proapo AI associated mainly with PE (56 %) and lyso-PE (17 %) and unidentified Px (19 %), the less dense particles 1.10-1.16 g/ml with proapo AII and exclusively lyso-PE (92 %) as newly synthesized lipid (figure 10 E).

These experiments raise the question whether the expression of an apolipoprotein gene, as demonstrated here for apo AI and AII, induce a specific synthesis of phospholipid classes for the formation of a supra-molecular structure for secretion, apolipoprotein and/or cell specific.

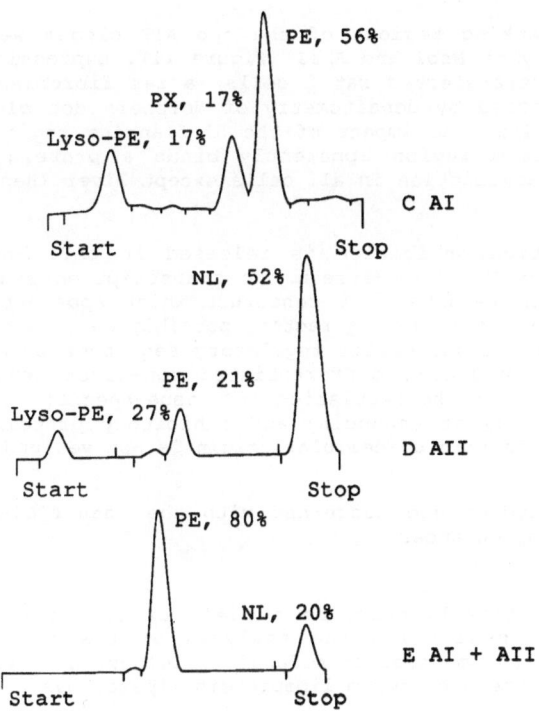

Fig. 10. Radio thin layer scans of *Xenopus laevis* oocyte lipids. Oocytes were injected with the DNA listed at the right.

TRANSIENT AND PERMANENT EXPRESSION OF APO AI AND AII
IN CHO AND RAT 2 (FIBROBLAST) CELL LINES

For the study of gene regulation of the main HDL apoproteins AI and AII, not only of the single gene but also of the interdependence of apo CIII and apo AI gene expression we have established permanently expressing CHO cell lines which are dehydrofolate reductase (Dhfr) negative (auxotroph for thymidine, hypoxanthine and glycine) with the calcium phosphate procedure and cotransfection with pCVSV-Dhfr cDNA as marker for transformed cells. 80 % of the clones contained both DNA species and were selected in methotrexate-supplemented selection medium. Northern blot analysis of CHO cell apo AI- and apo AII-specific RNA of CHO cells confirmed the expression.

The Dhfr gene is a reliable marker for apo AI-transfected CHO cells and RNA dot blots of the clones, but particular ELISA of the serum-free medium of the cell clones were strongly positive in a dilution of 1:100 as compared to the control medium.

The apo AI gene was also stably integrated after a period of two months.

Furthermore these cell lines allow experiments as described here for the apo AII gene regulation of expression.

The 5'-flanking regions of the apo AII clones were truncated by restriction enzymes *NcoI* and *ApaI* (figure 11). Expression of these DNAs in transiently transfected rat 2 cells, a rat fibroblast cell line and CHO cells, measured by densitometry of Northern dot blot hybridization analyses, underline the impact of the 5'-flanking region: λAII with the >1100 bp upstream region apparently binds a protein which inhibits splicing and transcription in all cells except liver (Hep G2) and enterocyte.

Transcription inhibition is released in pAII and AII *NcoI*. The latter still has 240 bp upstream the transcription start. Deletion by *ApaI* restriction results in a construct which apparently has lost the binding site for a stimulating factor, possibly an enhancer. This and the promoter are tissue-unspecific regulatory sequences because of thex are also expressed in little differentiated non-liver cells. Figure 11 B resembles a model of the regulation of tissue-specific expression of apo AII by the binding of enhancing and inhibitory proteins. The role of transactivation factors of cellular origin is not yet understood.

Present studies are concerned with the insufficiently understood properties of the enhancer.

3) Let me finally describe a field of our interest, namely the apolipoprotein design for the analysis of the *structure-function relationship by the methods of molecular biology*. As a paradigm *apo CII*, the activator protein of serum lipoprotein lipase, was studied.

An *EcoRI* insert of a λgt11 apo CII cDNA clone was isolated. Nucleotide sequence analysis indicated that 44 bp were missing at the 5' end of the coding region. The apo CII clone, 442 bp of full length, was constructed with synthetic oligonucleotides, shown schematically in figure 12, and cloned into pUC13.

For expression studies and the analysis of the translation products antibodies against human serum VLDL apo CII were raised and likewise against a synthetic peptide embracing Asp^7 to Thr^{16} (KLH conjugate). The antiserum titer was determined with the ELISA.

In vitro transcription of the apo CII cDNA cloned into the *BamHI/PstI* site of the pSP19 vector yielded a 0.4 kb transcript, as shown in the Northern blot with CII 48^{mer} as probe, figure 13.

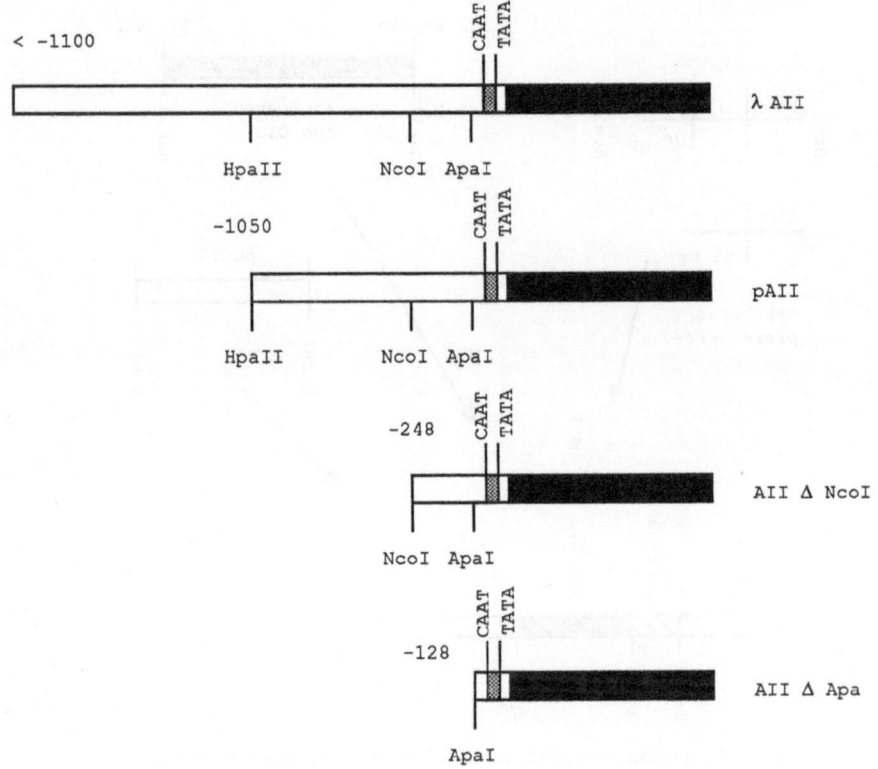

Fig. 11 A. 5' deletion mutants of genomic apo AII DNA.

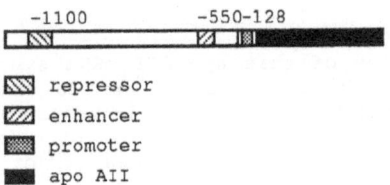

☒ repressor
▨ enhancer
▨ promoter
■ apo AII

Fig. 11 B. Proposed regulatory elements of 5'-untranslated region of human apo AII gene.

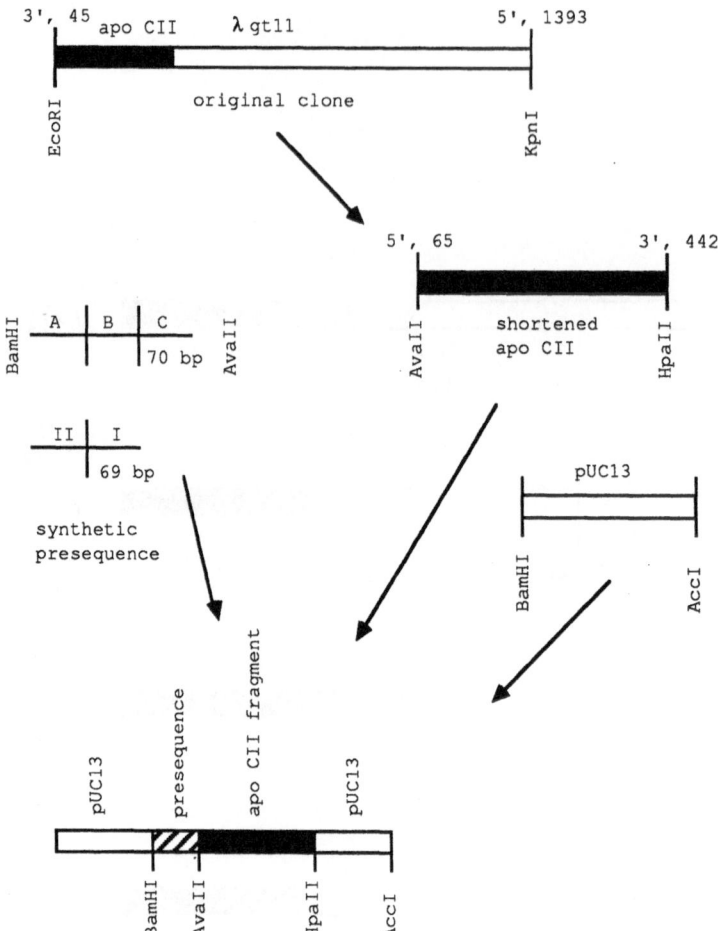

Fig. 12. Human apo CII cDNA cloned into pUC13 vector.

In vitro translation of this apo CII mRNA synthesized by the SP6
RNA polymerase, priming the reticulocyte or the wheat germ system and
purification of the translation product by immunoprecipitation with human
apo CII-specific antibodies, yielded a 11 kDa protein which is the pre-
form of apo CII with 101 amino acid residues (22 of the pre- and 79 of
the mature sequence).

For the production of nonfusion proteins of cloned eucaryotic DNA
in bacteria the apo CII cDNA was cloned into the pKK233-2 expression vector, figure 14. Sufficient amount of wild type and mutant apo CII to be
described later could be isolated from lysed bacteria with anti apo CII
IgG, purified by apo CII-peptide affigel 10 affinity chromatography. The
medium was free of apo CII. Apo CII, however, has been isolated from the
lysed bacteria. It was present as processed, 9 kDa large mature apo CII,
figure 15.

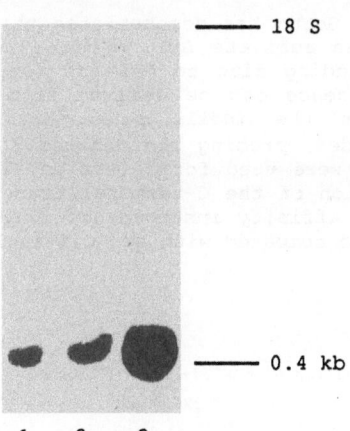

Fig. 13. Northern blot hybridization of 1 human liver mRNA (OD 40); 2 human liver mRNA (OD 32); 3 apo CII transcript from pSP19-CII clone.
Probe: apo CII-specific 38^{mer}.

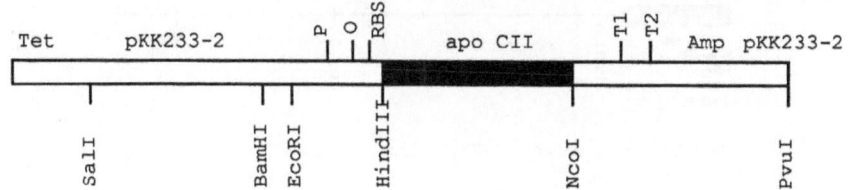

Fig. 14. Human apo CII cDNA cloned into the pKK233-2 vector.

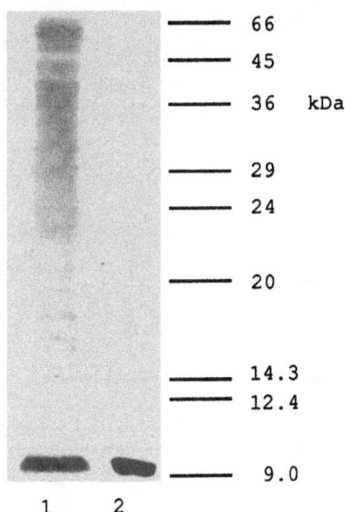

Fig. 15. SDS polyacrylamide gel electrophoresis of apo CII isolated from lysed E. coli transformed pKK233-2 apo CII: a) pulse-chase experiment; b) immunoprecipitated product from lysed cells.

27

The laboratory of Gotto has demonstrated that synthetic C-terminal peptides of apo CII can activate LpL. Other structural and functional domains such as the binding site to LPL and for fatty acids have been postulated. Direct evidence can be derived from 3'-terminal deletions with *Bal*31 exonuclease of the *Hin*dIII opened pKK233 apo CII clone, figure 16. 24[mer] oligonucleotides, probing the desired 3'-terminal deletions in six different mutants, were used for selection. These mutants were used for preparative isolation of the C-terminal truncated apo CII, isolated by the above described affinity immunoadsorption. They were assayed for their LPL activation and compared with apo CII isolated from VLDL.

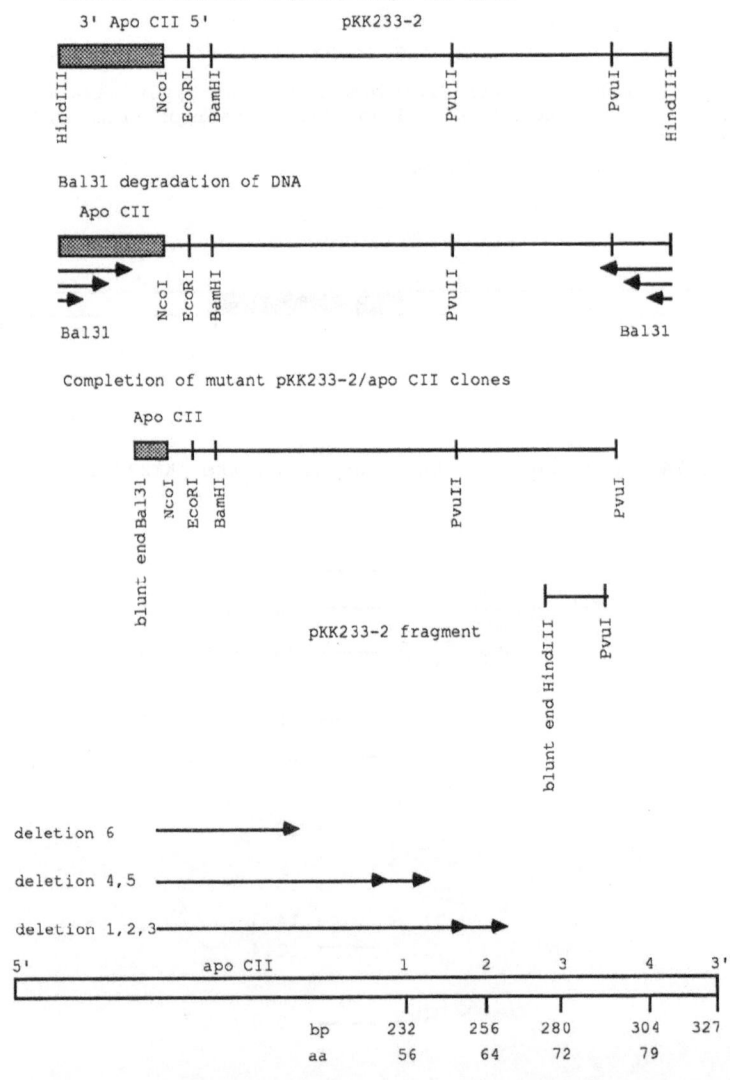

Fig. 16. Construction of deletion mutants of pKK233-2 apo AII with 3'-terminal truncated coding sequences obtained by partial *Bal*31 exonuclease digestion.

Table 1 clearly demonstrates that the deletion of only six C-terminal amino acid residues (73-79) reduces the activation of LPL to one third of that of the complete apo CII, and deletion by 14 residues yields an expression product which no longer enhances LPL-activity (table 1).

Activator	nM fFA/ml/h*	factor
apo CII of VLDL	649	6,4
apo CII (purified) of bacteria	601	5.8
apo CII del. 1	209	2.1
apo CII del. 2	236	2.3
apo CII del. 3	179	1.8
apo CII del. 4	147	
apo CII del. 5	161	
apo CII del. 6	161	
pKK233-2 wild type lysate (100-400 µl)	75	
test without activator	102	

*nM free fatty acids released from triolein/hour

CONCLUSION

I have tried to demonstrate the potential of recombinant DNA techniques in the studies of structural and functional aspects of lipoprotein research, exemplified in the field of processing and secretion of apo AI, regulation of expression of apo AI and AII in transfected eucaryotic cells and finally in the field of apolipoprotein engineering as exemplified for apo CII constructs and their potential for the activation of serum lipoprotein lipase.

Future studies open our common field of interest to unexpected and unbelievable horizons.

REFERENCES

Colman, A., 1985, in *Transcription and translation: a practical approach*, B. D. Harmes, and S. J. Higgins, eds.), IRL Press, Oxford.
Folz, R. J. and Gordon, J. I., 1986, *J. Biol. Chem.*, **261**: 14752.
Gurdon, J. B. and Melton, D. A., 1981, *Ann. Rev. Genet.*, **15**: 189.
Inouye, S., Wang, S., Sekizawa, J., Halegoua, S. and Inouye, M., 1977, *Proc. Natl. Acad. Sci. USA*, **74**: 1004.
Karathanasis, S. K., 1985, *Proc. Natl. Acad. Sci. USA*, **82**: 6374.
Kramer, W., Drutsa, V., Jansen, H. W., Kramer, B., Pflugfelder, H. and Fritz, H. J., 1984, *Nucl. Acids Res.*, **12**: 9441.
Smith, L. C., Voyta, J. C., Catapano, A. L., Kinnunen, P. K. J., Gotto, A. M. and Sparrow, J. I., 1980, *N.Y. Acad. Sci.*, **2**: 213.
Stoffel, W., Blobel, G. and Walter, P., 1981, *Eur. J. Biochem.*, **120**: 519.
Talmage, K., Stahl, S. and Gilbert, W., 1980, *Proc. Natl. Acad. Sci. USA*, **77**: 3369.

CIS-ACTING ELEMENTS AND TRANS-ACTING FACTORS INVOLVED IN CELL TYPE SPECIFIC EXPRESSION OF THE HUMAN APOLIPROTEIN AI GENE

U.Seedorf[+], K.N. Sastry[++], and S.K. Karanthanasis[++]

[+] Institut fur Arterioskleroseforschung, Universitat of Munster

[++] The Children Hospital, Department of Cardiology, Harvard Medical School, Boston

In mammals, the gene coding for apolipoprotein AI (apo AI) is expressed predominantly in the liver and the intestine (1). It is also expressed in the human hepatoma cell lines HepG2 and Hep3B, and in the human colon carcinoma cell line Caco2 (2). It is not expressed in HeLa cells. In this presentation we address the question, which DNA sequences of the human apo AI gene are responsible for its observed cell type specific expression. In addition, we investigated, if factors present in the nucleus of these cell types may be involved in the cell type specific expression of the human apo AI gene.

Cis-acting elements controll the cell type specific expression of the human apo AI gene

To map the 5´boundary of DNA sequences that are important for high level expression of the gene, a series of plasmids containing various lengths of DNA sequences flanking the 5´ end of the human apo AI gene were constructed and assayed for transient expression after introduction into cultured human hepatoma (HepG2), colon carcinoma (Caco2), and epithelial carcinoma (HeLa) cells (fig. 1a). The results showed that while most of these constructs are expressed in HepG2 and Caco2 cells, none of them is expressed in HeLa cells (fig. 1b). In addition, the results indicated, that a DNA segment located between nucleotides -253 and -38 upstream from the transcrip-

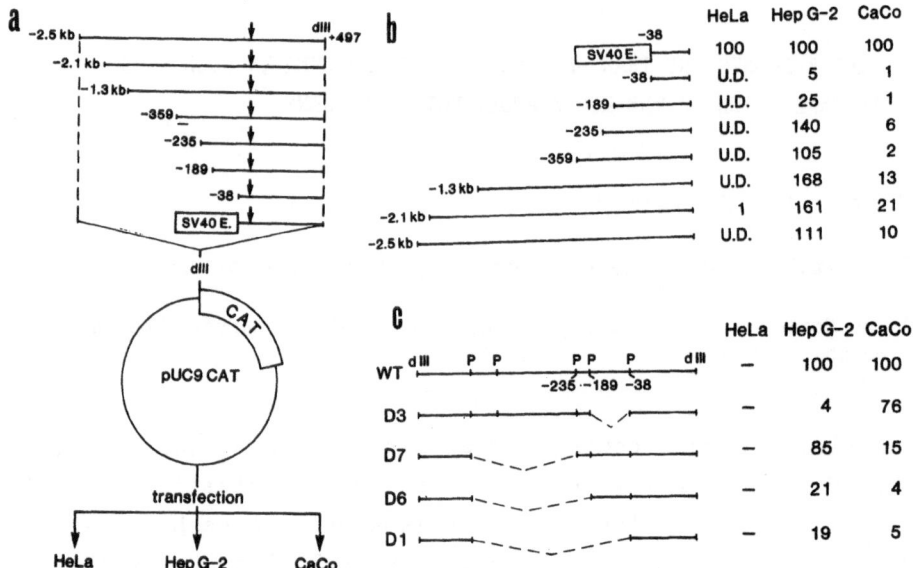

Fig. 1 Construction of apo AI-CAT (Chloramphenicol-Acetyltrans-
ferase) expression vectors (a).

The app. 3Kb long apo AI Hind III fragment, which contains
the start site of transcription, was cloned in the Hind III
site of the vector pUC9CAT (-3kbCAT.AI). Using additional
restriction endonucleases 5´deletions of the 3Kb apo AI
Hind III fragment were constructed as indicated and transfected into HeLa, HepG2, and Caco2 cells.

Expression of apo AI-CAT expression vectors containing
different 5´deletions (b) and various internal deletions
(c) after transfection into HeLa, HepG2, and Caco2 cells.

Values represent percentages of the values obtained with
the vector SV40CAT, which contained the SV40 enhancer at
position -38 of the human apo AI gene for (b) and the
wild type (WT) 3Kb apo AI Hind III fragment without internal deletion for (c). For methodological details see
(2).

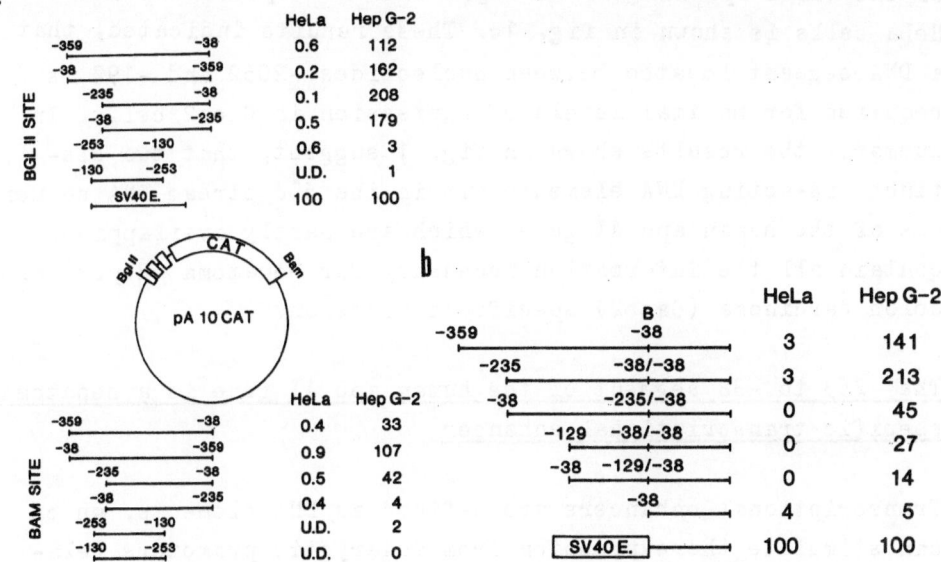

Fig. 2 Effects of position and orientation of apo AI 5′up-
stream DNA fragments on the stimulation of expression
of CAT activity under the control of the SV40 early
promoter in the vector pA10CAT (a).

The indicated fragments of the human apo AI gene
were cloned in A- and B-orientation in the Bgl II
and BamH1 restriction sites of the vector pA10CAT.
Expression was measured after transfection into HeLa
and HepG2 cells. The values represent percentages of
the values obtained with the vector pSV2CAT. For
details see reference 2.

Effect of the orientation of 5′upstream DNA sequences of the human apo AI gene on the expression of
apo AI-CAT fusion constructs in HeLa and HepG2 cells.

The Pst I site at position -38 of the human apo AI
gene was changed to a Bam H1 site and apo AI 5′up-
stream DNA fragments, which were converted to Bam H1
fragments by appropiate techniques, were cloned in
A- and B-orientation in this Bam H1 site and CAT expression was measured after transfection into HeLa
and HepG2 cells. The values represent percentages
of the values obtained with the construct containing
the SV40 enhancer (A-orientation) at this position.

tional start site of this gene is necessary and sufficient for
maximal levels of expression in HepG2 but not in Caco2 cells.
The effect of internal deletions within the 5'upstream region
of the human apo AI gene on expression in HepG2, Caco2, and
HeLa cells is shown in fig. 1c. These results indicated, that
a DNA segment located between nucleotides -2052 and -192 is
required for maximal levels of expression in Caco2 cells. In
summary, the results shown in fig. 1 suggest, that two distinct cis-acting DNA elements within the 5'upstream DNA sequences of the human apo AI gene, which are partly overlapping,
contain all the information necessary for hepatoma (HepG2) and
colon carcinoma (Caco2) specific expression.

The -253 to -38 segment of the human apo AI gene is a hepatoma specific transcriptional enhancer

Transcriptional enhancers are defined as DNA elements, which
can stimulate the expression from eukaryotic promoters relatively independent of the orientation and the distance with
respect to their target promoters (3). In fig. 2 it was shown
that this is the case for the -253 to -38 DNA segment when
inserted into two different positions and orientations in the
vector pA10CAT, which contains the Simian virus 40 (SV 40)
early promoter. When the orientation of the human apo AI
-253 to -38 DNA segment in front of the human apo AI gene
promoter was changed, as it was done in the experiment shown
in fig. 2b, this DNA fragment still retains most of its
stimulating activity and complete hepatoma specificity. In
addition, it was shown that the -253 to -38 DNA segment of the
human apo AI gene can stimulate the transcription from the
Adenindeaminase (ADA) promoter in a hepatoma specific fashion
(results not shown). Therefore, this DNA segment functions
as a hepatoma specific transcriptional enhancer with both,
homologous and heterologous promoters. These results indicated
that different cis- and possibly trans-acting factors are
involved in the establishment and subsequent regulation of
expression of the apo AI gene in the mammalian liver and
intestine.

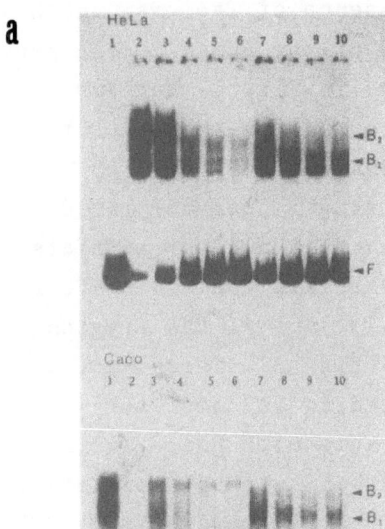

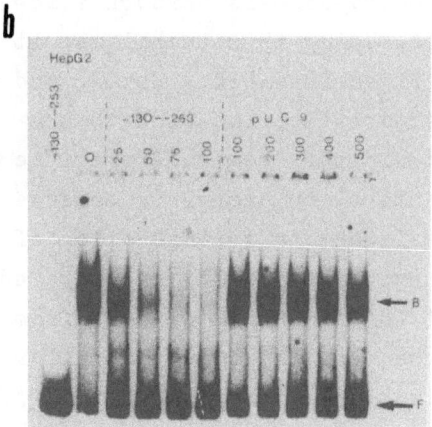

Fig. 3   Binding of nuclear factors derived from HeLa, HepG2, and
         Caco2 cells as indicated to the -253 to -130 apo AI
         enhancer fragment.

         HeLa (10ug), HepG2, and Caco2 (20ug) were prepared
         according to reference 8. Nuclear extracts were mixed
         with 0.5ng (32P)-endlabelled -253 to -130 DNA fragment
         (10,000cpm/ng) in the presence of 50 ng/ul poly(dI-dC)
         and bound (B) and free (F) DNA was separated by elec-
         trophoresis through a low ionic strength polyacrylamide
         gel. Binding ractions of lanes 2 to 6 contained excess
         of nonradiolabelled -253 to -130 DNA fragment, lanes
         7 to 10 contained pUC9 DNA as indicated for (b).

Nuclear extracts from HeLa, HepG2, and Caco2 cells contain proteins which bind to the -253 to -38 apo AI gene segment

For a variety of viral and cellular enhancers it was shown that enhancer action is mediated by trans-acting factors which can bind at specific binding sites within the enhancers (4-7). Therefore, we prepared nuclear extracts from HeLa, HepG2, and Caco2 cells and tested for the presence of factors, which can bind to the -253 to -38 DNA region of the human apo AI gene. Results of such an experiment are shown in fig. 3. Nuclear extracts were mixed with ^{32}P-endlabelled DNA fragments of the human apo AI gene and free and bound DNA was separated by electrophoresis through a low ionic strength polyacrylamide gel. Using this technique, it was shown that nuclear extracts derived from HeLa, HepG2, and Caco2 cells all contain nuclear factors that can bind to the -253 to -131 apo AI DNA fragment in a highly sequence specific fashion (fig. 3). These nuclear factors are proteins, because their binding is completely resistant to preincubation of the extracts with RNase, while they can be easily inactivated by treatment with proteinase K. If DNA fragments derived from the SV40 or Cytomegalovirus (CMV) enhancer, the long terminal repeats (LTR) of the Rous sarcoma virus (RSV) or from the apo AI-CIII-AIV gene locus were used for binding competition experiments, only fragments containing the -253 to -131 DNA fragment of the apo AI gene were able to compete effectively. The binding constant of specific versus nonspecific binding is 5.2 x 10^4 in case of the HeLa nuclear extract and 2.7 x 10^5 in the case of HepG2 nuclear extracts. Assuming 1:1 stochiometry of binding, a single HeLa cell contains app. 15,000 molecules, while a single HepG2 cell contains app. 4,000 molecules of this binding activity.

HeLa and HepG2 nuclear extracts contain a protein which binds to an imperfect octamer repeat sequence motif

To map the exact binding site of the proteins, DNaseI footprinting analyses using nuclear extracts of HeLa and HepG2 cells were performed. The results are shown in fig. 4a. The obtained footprint indicated that the binding site is 23 base pairs long and is located between nucleotides -191 and -213 from the start site. The binding site contains the imperfect octamer repeat sequence motif "TGAACCCT TGACCCCT". There is no

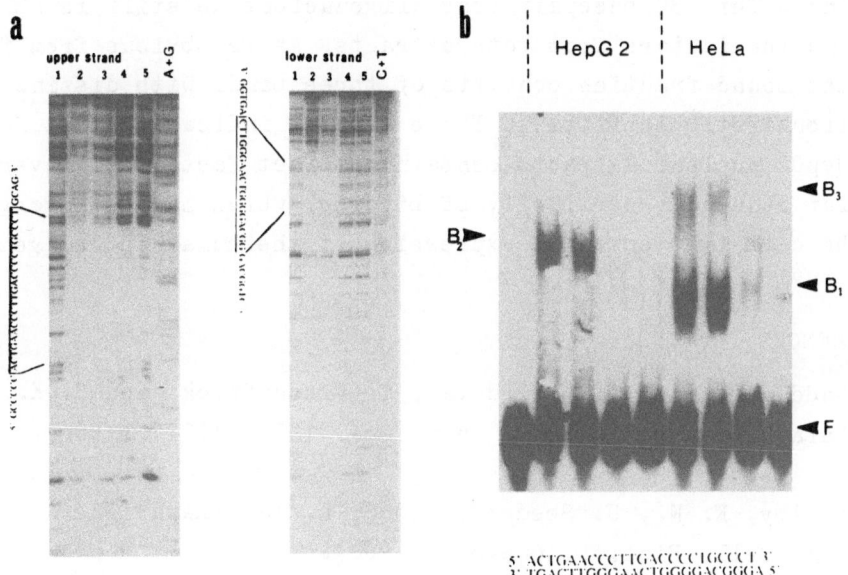

Fig. 4 DNase I footprinting after binding of HeLa (lanes 2, 4)
and HepG2 (lanes 3, 5) nuclear extracts to the -253 to
-130 apo AI enhancer fragment (a).

Nuclear extracts were prepared according to reference 8.
Binding was performed as described for figure 3, using
40ug (HeLa) and 80ug (HepG2) of extract protein and 2ng
of -253 to -130 DNA fragment labelled with (32P) at the
-253 (upper strand) or the -130 position (lower strand).
DNase I footprinting was performed as described in ref.
9.

Specific DNA binding of HeLa (10ug) and HepG2 (20ug)
nuclear extract protein to the synthetic double stranded
octamer repeat DNA (b).

Binding of the extracts was performed as described for
fig. 3. The sequence of the double stranded oligonucleotide was as indicated. Free DNA (F), bound DNA (B).

obvious homology between this DNA sequence and DNA sequences known to be important for the stimulation of generalized transcription. If a doublestranded synthetic oligonucleotide containing the octamer repeat sequence motif with some additional flanking nucleotides is used for binding of nuclear factors from HeLa and HepG2 cells, the results shown in fig. 4b are obtained. This 30 basepair long oligonucleotide still is able to bind the factors from both extracts. As is obvious from fig. 4b, the bound fraction consists of three bands with distinct electrophoretic mobilities. These results indicated, that HeLa and HepG2 nuclear extracts contain distinct factors with very similar sequence specificity of binding, which may be involved in the cell type specific expression of the human apo AI gene.

REFERENCES

1. Haddad, I. A., J. M. Ordovas, T. Fitzpatrick, and S. K.
   Karathanasis: J. Biol. Chem. 261, 13268-13277 (1986).

2. Sastry, K. N., U. Seedorf, and S. K. Karathanasis:
   Mol. Cell. Biol. 8, 605-614 (1988).

3. Khoury, G., and P. Gruss:
   Cell 33, 313-314 (1983).

4. Maniatis, T., S. Goodbourn, and J. A. Fischer:
   Science 236, 1237-1245 (1987).

5. Mercola, M., J. Goverman, C. Mirell, and K. Calame:
   Science 227, 266-270 (1985).

6. Dynan, W. S., and Tjian, R.:
   Nature 316, 774-778 (1985).

7. Becker, P. B., S. Ruppert, and G. Schütz:
   Cell 51, 435-443 (1987).

8. Dignam, J. D., R. M. Lebovitz, R. G. Roeder:
   Nuc. Acids Res. 11, 1475-1489 (1983).

9. Galas, D., and A. Schmitz:
   Nuc. Acids Res. 5, 3157-3170 (1978).

HUMAN APOLIPOPROTEIN A-I: STUDIES ON GENE EXPRESSION AND SITE-DIRECTED MUTAGENESIS IN *E. COLI*

Marco Soria*, Antonella Isacchi, Rolando Lorenzetti, Lucia Monaco, Raffaele Palomba, Paolo Sarmientos

Biotechnological Research, Farmitalia Carlo Erba, 24 Viale E. Bezzi, 20146 Milano, Italy and *School of Pharmacy, University of Milano, Italy

High Density Lipoproteins (HDL) and its major protein component, apolipoprotein A-I (apo A-I), play major roles in cholesterol metabolism. Several studies indicate that exogenous administration of HDL-like apolipoproteins might result in positive effects on various experimental models (Stein et al., 1979; Koizumi et al., 1986). However, the difficulty to obtain pure apo A-I from human plasma in sufficient amounts has severely limited the exploration of possible benefits deriving by the administration of this protein for therapeutic purposes, either alone or combined with different lipids. Therefore, production of apoA-I and of its variant forms by recombinant DNA techniques would be very useful to further pursue this approach.

Towards this goal, recombinant apo A-I was recently obtained in CHO cells, where it is secreted as a mature protein, the propeptide being cleaved by a hitherto unknown mechanism (Mallory et al., 1987). However, expression of the protein in prokaryotes like *E. coli* would allow large scale production at lower cost, and would permit production of the precursor protein proapo A-I, which is not obtainable in CHO cells. Production of naturally occurring molecular variants (Table I), such as the

TABLE 1. ApoA-I genetic variants naturally occurring in the human population

NAME	MODIFICATION	BIOCHEMICAL ALTERATIONS
MILANO	$Arg_{173} \rightarrow Cys$	FORMS DIMERS AND COMPLEXES WITH Apo AII
GIESSEN	$Pro_{143} \rightarrow Arg$	60-70 % OF NORMAL LCAT ACTIVATION ABILITY
MARBURG (MUNSTER 2)	$Lys107 \rightarrow 0$	40-60 % OF NORMAL LCAT ACTIVATION ABILITY
MUNSTER 3:		
FAMILY A	$Asp_{103} \rightarrow Asn$	
" B	$Pro_4 \rightarrow Arg$	NONE
" C	$Pro_3 \rightarrow His$	

'A-I Milano' variant (Franceschini et al., 1980; Weisgraber et al., 1980), or of newly engineered variant proteins or "muteins", could also be easily achieved by such systems.

Many problems were encountered when attempting to express apo A-I using *E. coli* phagemid vectors developed in our laboratory (Lorenzetti et al., 1985). One problem was the susceptibility of recombinant apo A-I to intracellular degradation during bacterial growth, with a half life of <10 min. as determined by pulse-chase experiments (Monaco et al., 1986; Mallory et al., 1987). Thus, expression of apo A-I in *E. coli* was achieved by stabilizing the gene product resorting to gene fusions. Initially, the apo A-I gene was fused to the 3' end of the gene coding for *E. coli* β-galactosidase (Monaco et al., 1985). This procedure successfully prolonged the half-life of the apo A-I fusion product. However, degradation of the apo A-I moiety still occurred in the fused protein, with an approximate half-life of about 45 min. (Lorenzetti et al., 1986).

Other reasons for poor expression of apo A-I in *E. coli* were further investigated by inserting, at the 5' end of the β-galactosidase gene, fragments derived from the 5' end of the gene coding for apo A-I. This procedure resulted in poor efficiency of initiation of mRNA translation, and low expression levels of the resulting protein. With the help of synthetic oligonucleotides, several random mutants were generated in the fragment deriving from the region coding for the NH2-terminal sequence of apo A-I. However, these mutants did not alter the resulting aminoacid sequence because of the degeneracy of the genetic code. Some mutations led to high levels of expression, underlying the importance of secondary structures at the 5' end of apo A-I mRNA, which influence the efficiency of translational initiation (Isacchi et al., 1987). As a result, conditions for efficient production of both apo A-I and proapo A-I in *E. coli* were identified (Isacchi et al., 1988).

The apo A-I gene was modified by site directed mutagenesis techniques to obtain the 'Milano' variant and other "muteins". All were efficiently expressed in *E. coli*, and the corresponding proteins could be detected by immunoassay with anti-human apo A-I antiserum (Lorenzetti et al., 1988).

HDL bind to a variety of cells and tissues, and their interactions with hepatocytes and macrophages via putative receptors have been described. However, it is still controversial if apo A-I is the recognition protein which directs the uptake of HDL into cells via a specific receptor. To investigate the direct role of apo A-I in this pathway, a fusion protein with staphylococcal protein A was studied by one of us (L. M.) in collaboration with K. Howell and R. Cortese at the European Molecular Biology Laboratory in Heidelberg. The interaction of the fused apo A-I - protein A with rat hepatocytes, and with human and mouse macrophages, permitted to follow the fate of the apo A-I moiety in these cells. The apo A-I - protein A was found to have very similar binding properties to HDL and to be efficiently internalized, indicating that even if deprived of its lipid bulk it could still interact with cells in a similar fashion as its natural counterpart (Monaco et al., 1987). An identical construction was obtained with apo A-I Milano fused to protein A, that in preliminary experiments exhibited similar binding characteristics to wild-type apo A-I - protein A (L. Monaco, unpublished results).

After cell breakage by standard methods like sonication, freeze-thawing or lysozyme treatment, the hybrid proteins obtained in E. coli after fusion to β-galactosidase could be easily purified by affinity chromatography on p-amino phenyl-β-D-thiogalactoside-Sepharose (Ullmann, 1984). Similarly, recombinant mature apo A-I and its mutagenized derivatives could be recovered from supernatants of E. coli extracts after cell breakage and purified on a polyclonal anti-apo A-I Affigel-10 affinity column (Lorenzetti et al., 1988). In contrast, the fused protein A - apo A-I could not be successfully extracted by analogous procedures, necessitating instead of a denaturation procedure of the sonicated bacteria by urea-alkali treatment, followed by renaturation after neutralization and dialysis (Monaco et al., 1987).

At the boundary between the "stabilizing" gene (i.e., β-galactosidase or protein A) and apo A-I, a synthetic oligonucleotide adaptor provided a sequence coding for a peptide substrate (Val-Asp-Asp-Asp-Asp-Lys) for the proteolytic enzyme enteropeptidase (enterokinase; Monaco et al., 1985; Monaco et al., 1987). Therefore, cleavage of the resulting fused gene product after affinity purification should, in principle, release mature apo A-I. However, optimizing recovery of recombinant apo A-I by this approach still requires further investigations (G. Orsini and L. Monaco, unpublished results).

In conclusion, the studies described in this article resulted in a variety of efficient approaches to optimize apo A-I gene expression in *E. coli*. In addition, both fused and mature proteins should constitute useful tools as reagents for diagnostic and therapeutic applications, giving the opportunity to investigate in more detail the biological and pharmacological role of these molecules.

REFERENCES

Franceschini, G., Sirtori, C.R., Capurso, A., Weisgraber, K.H. and Mahley, R.W., 1980, A-I Milano apoprotein: decreased HDL cholesterol levels with significant lipoprotein modifications and without clinical atherosclerosis in an Italian family, J. Clin. Invest. , 66: 892-900.

Isacchi, A., Palomba, R., Sarmientos, P. and M. Soria, 1987, Use of protein fusions to increase gene expression of recombinant apolipoprotein A-I, Protein Engineering. 1: 250

Isacchi, A., Sarmientos, P., Lorenzetti, R. and Soria, M., 1988, Mature apolipoprotein A-I and its precursor proapo A-I: influence of the sequence at the 5' end of the gene on the efficiency of expression in *E.coli* , Submitted

Koizumi, J. , Jadhav, A. , Okabayashi, K. and Thompson, G.R., 1986, Effect of Apo-HDL Phosphatidylcholine (PC) complexes on cholesterol efflux in WHHL rabbits, in Proc. of IX Int. Symp. on Drugs Affecting Lipid Metabolism, Florence,p. 72

Lorenzetti, R., Dani, M., Casati, M., Lappi, D.A., Martineau, D. Monaco, L., Shatzman, A., Rosenberg, M. and Soria, M., 1985, pFCE4, a new system of *E.. coli* expression-modification vectors, Gene , 39: 85-87

Lorenzetti, R., Sidoli, A., Palomba, R. ,Monaco, L., Martineau, D., Lappi, D.A. and Soria, M., 1986, Expression of the human Apolipoprotein A-I gene fused to the gene for *E. coli* beta-galactosidase, FEBS Letters, 194: 343-346

Lorenzetti, R., Monaco, L., Palomba, R., Isacchi, A., Sarmientos, P. and Soria, M., 1988, Human apolipoprotein A-I and variant forms of the same expressed in *E. coli,* European patent application , n. 87309318

Mallory, J.B. , Kushner, P.J. , Protter , A.A. , Cofer ,C.L. , Appleby, V.L., Lau, K. , Schilling , J.W. and Vigne , J.L., 1987, Expression and characterization of human apolipoprotein A-I in chinese hamster ovary cells, J. Biol. Chem., 262: 4241-4247.

Monaco, L., Lorenzetti, R.and Soria, M., 1985, Construction of an *E.coli* expression-modification vector for stabilization and rapid purification of labile proteins, Atti Assoc. Genetica Ital., 195 - 196

Monaco, L., Lorenzetti, R., Sidoli, A. and Soria, M., 1986, Versatile expression-modification vectors as tools for structure-function studies of Apolipoprotein A-I, in NATO ASI "Human apolipoprotein mutants: impact on atherosclerosis and longevity", C.R. Sirtori, A.V. Nichols & G. Franceschini, eds., Plenum Press, New York, 153-160

Monaco, L., Bond, H.M., Howell, K.E., and Cortese, R., 1987, A recombinant apo A-I - Protein A hybrid reproduces the binding parameters of HDL to its receptor, EMBO J. , 6:3253-3260.

Stein, O., Fainaru, M. and Stein, Y., 1979, The role of lysophosphatidyl choline and apolipoprotein A-I in the cholesterol removing capacity of lipoprotein deficient serum in tissue culture, Biochim. Biophys. Acta, 574: 495-504.

Ullman, A., 1984, One-step purification of hybrid proteins which have β-galactosidase activity, Gene , 29: 27-31.

Weisgraber, K. H., Bersot, T., Mahley, R.W., Franceschini, G. and Sirtori, C.R., 1980, A-I Milano apoprotein: isolation and characterization of a cysteine-containing variant of the apoA-I protein from human HDL, J. Clin. Invest. , 66: 901-907.

APOLIPOPROTEIN AI-MILANO: MECHANISMS FOR THE ANTIATHEROGENIC POTENTIAL

Guido Franceschini, Laura Calabresi, Massimo Baio, Alex V. Nichols, and Cesare R. Sirtori

Center E. Grossi Paoletti, Institute of Pharmacological Sciences, University of Milano, Italy and Donner Laboratory, University of California at Berkeley, USA

INTRODUCTION

The apolipoprotein AI-Milano (AI-M) has been the first described molecular variant of human apolipoproteins (1,2). It is characterized by a cysteine for arginine substitution at the position 173 in the primary sequence of apo AI (3), leading to the formation of disulphide bonded homodimers AI-M/AI-M and heterodimers with apolipoprotein AII (AI-M/AII) (2).

After the discovery of the AI-Milano mutant, several other molecular variants of apo AI have been identified through the world (4). None of these is associated with significant clinical-biochemical abnormalities in the affected carriers, although two, i.e. the AI-Marburg and AI-Giessen, are defective in the in vitro activation of the lecithin:cholesterol acyl transferase (LCAT) enzyme (4). By contrast, the AI-Milano mutant leads to marked alterations of the lipid-lipoprotein patterns in the affected subjects, most evident being a dramatic reduction of the high density lipoprotein (HDL) cholesterol and apo AI levels (1,5). In spite of this potentially harmful biochemical defect, the prevalence of atherosclerotic complications is unusually low in the carriers (6), suggesting that the AI-Milano mutant may offer some selective advantage against atherosclerosis development.

To investigate possible mechanisms responsible for the protective effect of the AI-Milano mutant, studies have been focused on the structural characterization of the AI-Milano molecule and on the functionality of the HDL system in the carriers.

STRUCTURAL PROPERTIES OF THE AI-MILANO MUTANT

The effect of the cysteine for arginine substitution in the monomeric AI-Milano variant on the structural properties of the AI molecule has been investigated by circular dichroism and spectrofluoroscopy (7). A 15% lower content of α-helix has been detected in the mutant, which also shows a 2 nm red shift in the wavelength maximum of the emission spectrum, when compared to normal apo AI. These experimental findings are consistent with the loss of one of the several amphipathic domains identified in the carboxy terminal region of the AI molecule. Theoretically, this may result from the disappearance of one of the two ion pairs (8), present in topographically close relationships in the amphipathic AI fragment between ser-167 and asn-184 (9), consequent to the substitution of the involved basic arginine residue in position 173 with a neutral cysteine in the mutant apolipoprotein. This relatively minor structural alteration significantly affects the lipid binding properties of the AI-Milano variant. These have been tested using two different synthetic phospholipids, dimyristoylphosphatidylcholine and dipalmitoylphosphatidylcholine (7,10). A 70% faster association rate was detected after incubation of the phospholipid with apo AI-Milano compared to normal AI, indicative of a higher kinetic affinity of the mutant for lipids. Furthermore, phospholipid-apolipoprotein complexes containing apo AI-Milano were more easily destroyed by denaturing agents, suggesting that the mutant is less tightly bound to lipids than normal AI.

The increased flexibility of the mutant apolipoprotein in the interaction with model phospholipids may have physiological implications. Monomeric AI-Milano can move more easily from one lipoprotein particle to another, or dispose more efficiently of the associated lipids following interaction with cells (11). Both of these mechanisms, especially when associated with the accelerated in vivo catabolism of the mutant apolipoprotein (12), can contribute to an overall increased activity of the AI-Milano molecule in tissue lipid removal.

In addition to lipid binding, apolipoprotein AI plays other major roles in lipoprotein metabolism, being the physiological activator of LCAT (13) and acting as signal for the binding of AI-containing lipoproteins to the putative HDL receptor (14). Two apo AI mutants, AI-Giessen (Pro-143 ⟶ Arg) and AI-Marburg (Lys-107 ⟶ 0) are deficient in the activation of LCAT in vitro (15,16), because of a perturbation in the secondary structure of the AI molecule. The aminoacid replacement and the consequent structural alteration in the AI-Milano variant have no effect on the ability of the monomeric mutant in stimulating cholesterol esterification. However, when incorporated into model lipoproteins, both the AI-Milano homo- and hetero-dimers are totally defective in LCAT activation (4). Since all the identified AI-Milano carriers are heterozygotes for the presence of the mutant and very small concentrations of the functional AI are sufficient for LCAT activation (13), it is unlikely that the defective dimers are responsible for the overall decreased cholesterol esterification in the affected subjects (5). However, as the LCAT reaction plays a central role in the regulation of the HDL system, the AI-Milano dimers may affect in some way both the structure and the functionality of HDL particles. In any case, these findings will contribute to the understanding of the structural requirements necessary for LCAT activation by apolipoproteins. Specific aminoacid sequences or conformational properties have been potentially involved in this process, in addition to the physical structure of the substrate lipid-apolipoprotein particles (17). The future availability of purified dimers will allow us to correlate their structural and lipid-binding properties with the relative ability to activate LCAT.

THE HDL SYSTEM IN THE AI-MILANO CARRIERS

The most evident biochemical abnormality in the AI-Milano carriers is the already mentioned marked reduction of plasma HDL levels. All the HDL constituents, both lipids and apolipoproteins, are reduced, suggesting that the number of circulating HDL particles is lower, compared to controls (5). Furthermore, the ultracentrifugal HDL fraction (d=1.063-1.21

g/ml) purified from the AI-Milano subjects contains more protein and triglyceride and less cholesteryl esters tha normal (5). This reflects alterations both in the distribution of HDL subfractions and in the lipoprotein structure. When initially analyzed by the cross-linking technique, the large HDL-2 particles appeared structurally normal, although markedly reduced in number (18). The most prominent abnormality was in the HDL-3 subfraction, consisting of different particle subpopulations instead of the single one detected in controls. These studies have been extended to almost all the AI-Milano carriers (34, up to now) by the use of rate zonal ultracentrifugation and nondenaturing polyacrylamide gradient gel electrophoresis (GGE), thus allowing a complete quantitative characterization of HDL particle distribution and composition (Table) (19). The mean plasma HDL-2 mass and cholesterol levels are 67% and 80% lower in the AI-Milano carriers compared to controls, confirming the cross-linking data, and reflecting both a decrease in the number of particles, also apparent from the inspection of the GGE profiles (Figure), and a selective depletion of cholesterol, particularly cholesteryl esters. These are only partially substituted with triglycerides in the lipoprotein core, leading to the formation of smaller HDL-2 particles. Similar compositional changes are also detected in the HDL-3 subfraction, particles being triglyceride and protein enriched and cholesteryl esters poor. The HDL-3 mass and cholesterol are reduced by 40% and 50%, respectively, compared to controls. The flotation rate of HDL-3 is lower in the AI-Milano carriers, reflecting both the low lipid to protein ratio and the presence of different small and dense particle subpopulations in this density fraction.

Table I. Plasma levels, structure and compostition of HDL particles in AI-Milano carriers and in age-sex matched controls.

	AI-Milano	Controls
HDL-2		
Mass (mg/dl)	29.6 ± 10.4	85.9 ± 46.7
Ve (ml)	7.0 ± 0.2	7.1 ± 0.2
Size (HDL-2b, nm)	9.9 ± 0.2	10.3 ± 0.3
FC%	1.9 ± 1.3	3.4 ± 1.2
CE%	17.0 ± 1.9	25.2 ± 2.8
TG%	5.9 ± 1.5	3.5 ± 1.1
PL%	27.2 ± 3.5	26.4 ± 3.9
P%	47.8 ± 5.7	41.5 ± 4.7
HDL-3		
Mass (mg/dl)	151.8 ± 57.8	254.0 ± 58.3
Ve (ml)	3.3 ± 0.4	4.0 ± 0.2
Size (HDL-3a, nm)	8.6 ± 0.1	8.4 ± 0.1
Size (HDL-3b, nm)	8.0 ± 0.1	7.9 ± 0.1
FC%	2.2 ± 0.6	2.2 ± 0.2
CE%	13.1 ± 2.9	20.9 ± 2.6
TG%	5.0 ± 1.4	2.9 ± 1.0
PL%	26.6 ± 2.4	25.6 ± 2.5
P%	53.1 ± 3.4	48.3 ± 4.7

Ve: volume of elution from the rate zonal gradient; FC: free cholesterol; CE: cholesteryl esters; TG: triglycerides; PL: phospholipids; P:proteins.

As already noted in the cross-linking studies (18), the GGE separations of AI-Milano HDL revealed the presence of two major peaks with maxima within the particle intervals of HDL-3 subpopulations, namely HDL-3a and HDL-3b (Figure). Compared to control HDL, the AI-Milano HDL are unique in consistently exhibiting a distinct peak within the HDL-3b size interval. However, a great variability in the realtive contribution of the two HDL-3 peaks to the total HDL profile can be observed among the various AI-Milano subjects. Three patterns can be distinguished: pattern I is characterized by a relative abundance of the smaller HDL-3b; in pattern II, the contribution of HDL-3a and HDL-3b is nearly equivalent, whereas HDL-3a particles predominate in pattern III (Figure). These qualitative differences can be quantified by integrating the areas below the GGE

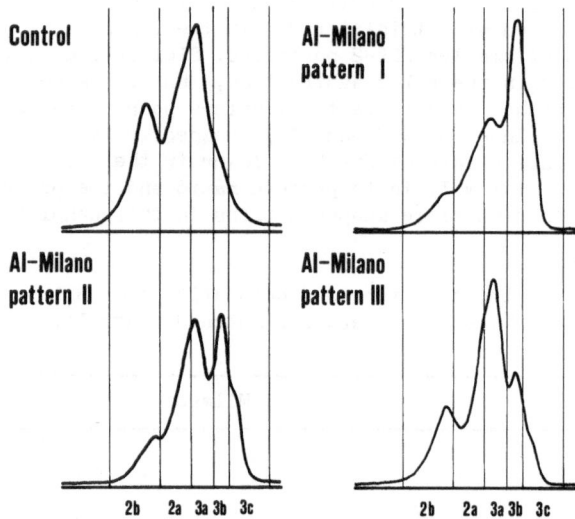

Fig. 1. Gradient gel electrophoretic profiles of high density lipoproteins isolated from a control subject and from carriers of the apo AI-Milano mutant showing different HDL patterns.

peaks. The calculated mean HDL-3a/HDL-3b mass ratio in the eight subjects with pattern I is 0.73 ± 0.10 and increases to 1.04 ± 0.08 and to 1.56 ± 0.22 in carriers with pattern II (n=7) and III (n=9), respectively. By analysis of plasma lipid levels in the examined subjects, a progressive, relative enrichment of HDL patterns with species of smaller particle size as a function of decreasing HDL-cholesterol and increasing triglyceride levels can be observed. Subjects with pattern I have the lowest HDL-cholesterol (12 ± 4 mg/dl) and the highest triglyceride (197 ± 87 mg/dl) values, whereas carriers with pattern III have normal triglyceride (112 ± 34 mg/dl) and subnormal HDL-cholesterol (26 ± 6 mg/dl) levels. Furthermore, pattern I subjects show a relative abundance of AI-Milano compared to AI, whereas the AI-Milano to AI ratio is about 1:1 in carriers with pattern III. These data indicate that a complex alteration in the lipoprotein system, linking together the amount of mutant apolipoprotein and both triglyceride and HDL metabolisms, is present in the AI-Milano carriers, resulting in anomalous structural features of HDL particles. These are associated with major functional alterations in the HDL system; in a subset of affected and unaffected subjects, both the cholesterol esterification and the CETP mediated transfer of newly formed cholesteryl esters from HDL to apo B containing lipoproteins are reduced in the carriers (16.3 and 7.2 µg/ml/hr, respectively) compared to controls (22.0 and 18.8 µg/ml/hr).

CONCLUSIONS

In an attempt to disclose possible mechanisms responsible for the anti-atherogenic potential of the AI-Milano mutant, several experiments have been carried out, focusing on the structural features of the mutant apolipoprotein, on the functionality of its molecular forms and on the

characterization of the HDL system in the carriers. These studies show that the monomeric AI-Milano is a hyperfunctional form of apo AI, whereas its dimers are functionally defective.

The lipoprotein system, including both triglyceride and HDL metabolisms, is markedly altered in the AI-Milano carriers, with an apparent "dosage effect" of the mutant in dictating the severity of the alterations. This variability in the phenotypic expression of the AI-Milano syndrome may depend from the coexistence in the plasma of a functionally normal AI apolipoprotein, of the hyperfunctional monomeric AI-Milano and of the defective AI-Milano dimers. Future in vitro and in vivo experiments using purified apolipoproteins and model lipoproteins will be of help in the understanding of the mechanisms involved in the regulation of the complex balance between these different apolipoprotein forms in the individual. Since the efficiency of the involved apolipoproteins plays a major role in the general functionality of the HDL system and thus of the reverse cholesterol transport processes, these studies will also identify the molecular mechanisms protecting the AI-Milano carriers from atherosclerosis development.

Supported by the Consiglio Nazionale delle Ricerche of Italy (PF Ingegneria Genetica e Basi Molecolari delle Malattie Ereditarie).

REFERENCES

1. G. Franceschini, C.R. Sirtori, A. Capurso, K.H. Weisgraber and R.W. Mahley, AI-Milano apoprotein. Decreased high density lipoprotein cholesterol levels with significant lipoprotein modifications and without clinical atherosclerosis in an Italian family, J. Clin. Invest., 66: 892 (1980).
2. K.H. Weisgraber, T.P. Bersot, R.W. Mahley, G. Franceschini and C.R. Sirtori, AI-Milano apoprotein. Isolation and characterization of a cysteine-containing variant of the AI apoprotein from human high density lipoproteins, J. Clin. Invest., 66: 901 (1980).
3. K.H. Weisgraber, S.C. Rall Jr, T.P. Bersot, R.W. Mahley, G. Franceschini and C.R. Sirtori, Apolipoprotein AI-Milano. Detection of normal AI in affected subjects and evidence for a cysteine for arginine substitution in the variant AI, J. Biol. Chem., 257: 9926 (1983).
4. R.W. Mahley, T.L. Innerarity, S.C. Rall Jr and K.H. Weisgraber, Plasma lipoproteins: apolipoprotein structure and function, J. Lipid Res., 25: 1277 (1984).
5. G. Franceschini, C.R. Sirtori, E. Bosisio, V. Gualandri, G.B. Orsini, A.M. Mogavero and A. Capurso, Relationship of the phenotypic expression of the AI-Milano apoprotein with plasma lipid and lipoprotein levels, Atherosclerosis, 58: 159 (1985).

6. V. Gualandri, G. Franceschini, C.R. Sirtori, G. Gianfranceschi, G.B. Orsini, A. Cerrone and A. Menotti, AI-Milano apoprotein. Identification of the complete kindred and evidence of a dominant genetic transmission, Amer. J. Human. Gen., 37: 1083 (1985).

7. G. Franceschini, G. Vecchio, G. Gianfranceschi, D. Magani and C.R. Sirtori, Apolipoprotein AI-Milano. Accelerated binding and dissociation from lipids of a human apolipoprotein variant, J. Biol. Chem., 260: 16321 (1985).

8. A. Bierzynski, P.S. Kim and R.L. Baldwin, A salt bridge stabilizes the helix formed by isolated C-peptide of RNase A, Proc. Natl. Acad. Sci. USA, 79: 2470 (1982).

9. J.P. Segrest and R.J. Feldmann, Amphipathic helixes and plasma lipoproteins: a computer study, Biopolymers, 16: 2053 (1977).

10. G. Franceschini, L. Calabresi, P. Apebe, M. Sirtori, G. Vecchio and C.R. Sirtori, Apolipoprotein AI-Milano: a structural modification in an apolipoprotein variant leading to unusual lipid binding properties, in: Human apolipoprotein mutants: impact on atherosclerosis and longevity (C.R. Sirtori, A.V. Nichols and G. Franceschini, eds), Plenum Press, N.Y., p. 71 (1986).

11. H.B. Brewer Jr, G. Ghiselli, E.J. Shaefer, L.A. Zech, G. Franceschini and C.R. Sirtori, Apolipoprotein AI-Milano: in vivo metabolism of an apolipoprotein AI variant, in: Human apolipoprotein mutants: impact on atherosclerosis and longevity (C.R. Sirtori, A.V. Nichols and G. Franceschini, eds), Plenum Press, N.Y., p. 95 (1986).

12. C.J. Fielding, V.G. Shore and P.E. Fielding, A protein cofactor of lecithin:cholesterol acyltransferase, Biochem. Biophys. Res. Comm., 46: 1493 (1972).

13. N.H. Fidge, Partial purification of a high density lipoprotein-binding protein from rat liver and kidney membranes, FEBS Lett., 199: 265 (1986).

14. G. Utermann, J. Haas, A. Steinmetz, R. Paetzold, S.C. Rall Jr, K.H. Weisgraber and R.W. Mahley, Apolipoprotein AI-Giessen: a mutant that is defective in activating lecithin:cholesterol acyltransferase, Eur. J. Biochem., 144: 325 (1984).

15. S.C. Rall Jr, K.H. Weisgraber, R.W. Mahley, Y. Ogawa, C.J. Fielding, G. Utermann, J. Haas, A. Steinmetz, H.J. Menzel and G. Assmann, Abnormal lecithin:cholesterol acyltransferase activation by a human apolipoprotein AI variant in which a single lysine residue is deleted, J. Biol. Chem., 259: 10063 (1984).

16. A. Jonas, Synthetic substrates of lecithin:cholesterol acyltransferase, J. Lipid Res., 27: 689 (1986).

17. G. Franceschini, T.G. Frosi, C. Manzoni, G. Gianfranceschi and C.R. Sirtori, High density lipoprotein-3 heterogeneity in subjects with the apo AI-Milano variant, J. Biol. Chem., 257: 9926 (1982).

18. G. Franceschini, L. Calabresi, C. Tosi, C.R. Sirtori, C. Fragiacomo, G. Noseda, E. Gong, P. Blanche and A.V. Nichols, Apolipoprotein AI-Milano: correlation between high density lipoprotein subclass distribution and triglyceridemia, Arteriosclerosis, 7: 426 (1987).

IN VIVO CATABOLISM OF APOLIPOPROTEIN A-I IN SUBJECTS WITH FAMILIAL HYPOALPHALIPOPROTEINEMIA

Roma P., Gregg R.E., Meng M., Bishop C., Ronan R., Zech L.A., Meng M.V., Glueck C.*, Vergani C.▲, Franceschini G.●, Sirtori C.R.●, and Brewer H.B., Jr.

Molecular Disease Branch, NHLBI, NHI, Bethesda, MD.
*Cholesterol Center, Jewish Hospital, Cincinnati, OH.
▲Institute of Internal Medicine, University of Milano, Italy.
●Centro E.Grossi Paoletti, Dipartimento di Scienze Farmacologiche, Milano, Italy

Plasma levels of high density lipoproteins (HDL), specifically the concentration of HDL-cholesterol and apolipoprotein A-I, are inversely correlated with the incidence of coronary artery disease, (CAD) in humans (1-7). Clinical and epidemiological studies indicated that hypoalphalipoproteinemia and low plasma levels of apoA-I are risk factors for the development of cardiovascular disease.

ApoA-I, the major protein of HDL, is secreted as 249 amino acid long protein with subsequent cleavage in plasma to its 243 amino acid long mature form (8). The gene for apoA-I is located on the long arm of chromosome 11, and is part of a multigene complex including the genes for apoC-III and apoA-IV (9). In a number of kindreds, a low level of apoA-I in plasma is inherited as a genetically dominant trait. We have investigated the in vivo metabolism of apoA-I in two different types of dominant hypoalphalipoproteinemia, primary familial hypoalphalipoproteinemia (5-6) and hypoalphalipoproteinemia secondary to apoA-I Milano (10), in order to gain additional insights into the etiology of the hypoalphalipoproteinemia in these kindreds.

Primary familial hypoalphalipoproteinemia (5-6) is characterized by HDL cholesterol levels below the 10th percentile, with normal plasma concentrations of cholesterol and triglycerides and is linked to premature CAD. Ordovas et al. (7) observed that individuals with primary familial hypoalphalipoproteinemia had higher frequency of a specific restriction fragment length polymorphism (RFLP) of their DNA linked to the apoA-I gene compared to normal individuals.

Most normal individuals had 2.2 kb Pst1 restriction endonuclese fragment of their apoA-I gene DNA (11) while a majority of the primary familial hypoalphalipoproteinemia patients were either heterozygous or homozygous for 3.3 kb fragment (7, 12). Therefore, this polymorphism has been proposed to be a genetic marker for a mutation which causes hypoalphalipoproteinemia (7), with the apoA-I gene being the most likely mutant one.

We investigated the kinetics of apoA-I metabolism in three unrelated hypoalpha males who were either homozygous or heterozygous for 3.3 kb Pst I polymorphism (Table I).

TABLE I. Characteristics of Hypoalpha and Control Subjects.

Subject	(Yr)	Chol	TG	(mg/dl) HDL-CHOL	ApoA-I	KB PstI
Hypoalpha 1	23	156	203	28	100	3.3/3.3
Hypoalpha 2	32	169	126	33	96	2.2/2.2
Hypoalpha 3	40	240	437	26	109	3.3/3.3
Controls (n=6)	25±4*	173±30	78±20	54±15	127±18	-

* Mean ± S.D.

Hypoalpha 1 has never been described, hypoalpha 2 is subject III-6 in the family 14 described Vergani et al.(5), and Hypoalpha 3 is the proband of family 14 described by Third et al. (6).

HDL_2 and HDL_3 from each subject were isolated by sequential ultracentrifugation and their composition for free cholesterol, cholesterol esters, triglycerides, phospholipids and total protein were quantited. Both subclasses of HDL had a normal relative composition in the hypoalpha subjects, but the absolute concentrations of the components of HDL_2 were decreased in these individuals, while the level of HDL_3 was normal. This suggests that their hypoalphalipoproteinemia is the result of a decreased number of HDL_2 particles with normal composition. Amino acid analysis (13) and two dimensional (2-D) gel electrophoresis (14) were performed on apolipoprotein A-I isolated from the plasma of control and hypoalpha subjects. No abnormality was detected in the amino acid composition or in the 2-D pattern of hypoalpha apoA-I.

The metabolic behaviour of these was investigated by in vivo kinetic studies. Purified apoA-I from normal and hypoalpha subjects were iodinated with ^{131}I and ^{125}I respectively, by a modification of the IC1 method (15), reassociated with plasma lipoproteins, and both forms of apoA-I were simultaneously injected into normal and hypoalpha subjects. Plasma samples were obtained over time and the residence times (RT) of apoA-I were calculated as the area under the plasma decay curves (16). The residence time of apoA-I from normal and hypoalpha subjects were virtually identical in normal subjects and also in hypoalpha subjects. Hypoalpha subjects, however, catabolized autologous apoA-I at faster rate than normal controls (Figure 1).

Table II summarizes the kinetic parameters determined from the metabolic study.

ApoA-I is catabolized in the hypoalpha subjects at a faster rate than in normal subjects while the production rate

conclude that in these individuals hypoalphalipoproteinemia is due to the rapid catabolism of a structurally and metabolically normal apoA-I which is synthesized at a normal rate.

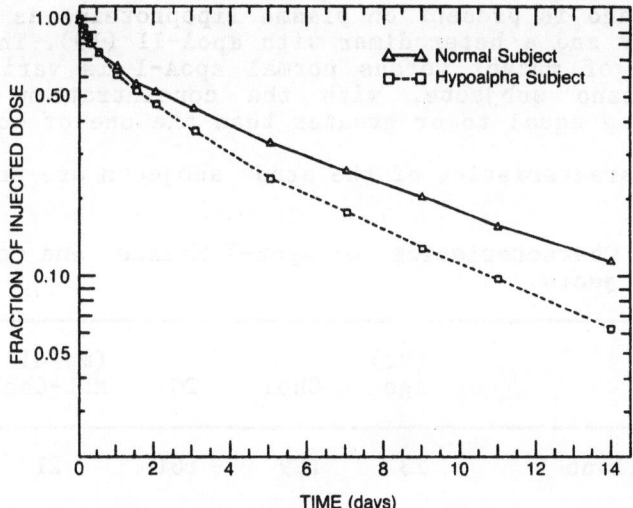

Figure 1. Plasma decay curves of normal apoA-I in a normal subjects and hypoalpha subjects.

TABLE II. Kinetic Parameters of ApoA-I Metabolism in Hypoalpha in Control Subjects.

Subjects	(mg/dl) ApoA-I	(days) RT*	(mg/Kg day) PR[+]
Hypoalpha 1	100	3.94	10.16
Hypoalpha 2	96	3.87	9.91
Hypoalpha 3	109	4.28	10.20
Controls (n=7)	126±17[&]	4.91±63**	10.28±1.23

*Residence Time
[+]Production Rate
[&]Mean ± S.D.
** p 0.01 compared to hypoalpha subjects.

Since both the structural and the regulatory regions of the apoA-I gene are normal we conclude that their apoA-I gene is normal. In addition, it is very likely that the PstI polymorphism linked to the apoA-I gene in familial hypoalpha subjects is a marker for a mutation that is close to, but not in, the apoA-I gene.

Apo-I metabolic studies were also performed in two individuals with hypoalphalipoproteinemia associated with a mutant form of apoA-I, apoA-I Milano. As previously described, the Milano variant of apoA-I is characterized by a cysteine

for arginine substitution at amino acid position 173 (17).
All the individuals with this mutation are heterozygous for it
and have marked hypoalphalipoproteinemia; there is an increase
of the incidence of hypertriglyceridemia in affected
individuals (18). Due to the presence of a cysteine residue,
apoA-I Milano is present on plasma liproteins as a monomer,
a homodimer and a heterodimer with apoA-II (19). The relative
proportion of mutant versus normal apoA-I is variable among
apoA-I Milano subjects, with the concentration of apoA-I
Milano being equal to or greater than the one of normal apoA-
I.

The characteristics of the study subjects are presented in
Table III.

TABLE III. Characteristics of ApoA-I Milano and Control Sub-
           jects.

Subjects	(Yr) Age	Chol	TG	(mg/dl) HDL-Chol	ApoA-I
ApoA-I Milano 1	23	199	181	21	86
ApoA-I Milano 2	33	169	166	12	51
Control 1	20	160	45	43	119
Control 2	21	130	43	47	120

The apoA-I Milano subjects had decreased plasma levels of
apoA-I and HDL cholesterol with mild hypertriglycedemia. The
patterns of apoA-I as obtained by two dimensional gel
electrophoresis (14) of reduced plasma from the apoA-I Milano
subjects and one normal control are illustrated in Figure 2.
Apo-I Milano has a one negative charge unit shift compared to
normal ApoA-I Milano; this results in two major mature
isoforms of apoA-I in the apoA-I Milano subjects. The normal
mature apoA-I is in the $A-I_0$ position while the mutant form is
in the $A-I_{-1}$ position. There is also a corresponding shift in
the position of the proform and of the minor mature forms of
apoA-I Milano.

In vivo metabolic studies of normal apoA-I were performed
as described earlier (15). ApoA-I Milano was purified from HDL
by Thiopropyl Sepharose 6-B affinity chromatography (17),
iodinated with ICl and reassocieted with plasma lipoproteins.
Lipoproteins containing ^{131}I-apoA-I and ^{125}I-apoA-I Milano
were simultaneously injected in two normal and two apoA-I
Milano subjects.

The kinetics of plasma metabolism of both types of apoA-I
were quantitated and the catabolism of apoA-I Milano monomer
and dimers was evaluated following their separation by SDS-
PAGE. In the control subjects apoA-I Milano was catabolized at
a faster rate than normal apoA-I. The apoA-I Milano subjects
catabolized both forms of apoA-I at a faster rate than the
control subjects, with the mutant form being catabolized
slightly faster than normal apoA-I in the first apoA-I Milano
subject and at the same rate in the second subject. The
synthetic rate of apoA-I was normal and low normal in the two
patients, respectively. From these data we conclude that

hypoalphalipoproteinemia in apoA-I Milano individuals is due to the rapid catabolism of a structurally and metabolically abnormal apoA-I, which is synthesized at a normal rate.

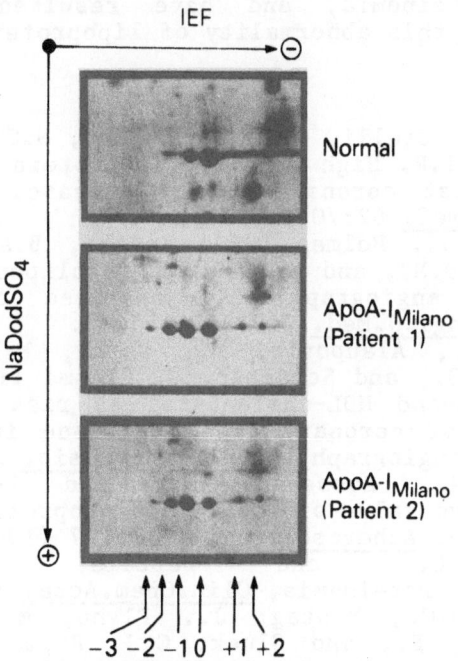

Two Dimensional Gel Electrophoresis of Plasma from Normal and ApoA-I$_{Milano}$ Subjects

IEF

NaDodSO$_4$

Normal

ApoA-I$_{Milano}$ (Patient 1)

ApoA-I$_{Milano}$ (Patient 2)

−3 −2 −1 0 +1 +2

Figure 2. Two-dimensional gel electrophoresis patterns of reduced apoA-I in total plasma from a normal control and the two apoA-I Milano subjects. These subjects are heterozygous for apoA-I Milano: the mutant form of apoA-I is present in the −1 position.

Evaluation of the decay curves of monomeric and dimeric apoA-I Milano demonstrated that apoA-I Milano monomers were catabolized at a much faster rate then the dimers, accounting for most of the catabolism of the mutant protein. The faster catabolism of monomeric apoA-I Milano results in an increased rate of catabolism of normal apoA-I which is catabolized on the same particle as apoA-I Milano. The dimers containing apoA-I Milano were catabolized more slowly than normal apoA-I. On the basis of these observations we conclude that the rate of formation of dimers containing apoA-I Milano in the different apoA-I Milano subjects is critical in determining the proportion of the normal versus mutant form of apoA-I in plasma, and in regulating the overall catabolic rate and plasma level of apoA-I in these subjects.

In conclusion, we have quantitated the kinetics of apoA-I metabolism in two types of subjects with familial hypoalphalipoproteinemia, and have demonstrated that the hypoalpha results from two very different mechanism. The first group of patients who had hypoalpha associated with Pst1 RFLP near the apoA-I gene, had low HDL and apoA-I levels secondary to the rapid catabolism of a metabolically normal apoA-I; therefore, the mutant gene resulting in familial hypoalpha in these subjects is not the gene for apoA-I. In the second class of subjects, patients with apoA-I Milano, the hypoalpha clearly resulted from the rapid catabolism of a structurally and metabolically abnormal apoA-I. Therefore, the gene for apoA-I is the site of the causative mutation for this syndrome. These studies have provided important new insights into the basic metabolic abnormalities leading to familial hypoalphalipoproteinemia, and have resulted in a better understanding of this abnormality of lipoprotein metabolism.

REFERENCES

1. Gordon, T., Castelli, W.P., Hjortland, M.C., Kannel, W.B., and Dawber, T.R. High density lipoprotein as a protective factor against coronary artery disease. The Framingham study, Am.J.Med. 62:707 (1977).
2. Maciejko, J.J., Holmes D.R., Kottke, B.A., Zimsmaister, A.R., Dinh, D.M., and Mao, S.J.T. Apolipoprotein A-I as a marker for angiographically assessed coronary artery disease. N.Engl.J.Med. 309:385 (1983).
3. Whayne, T.F., Alaupovic, P., Curry, M.D., Lee, E.T., Anderson, P.S., and Schecter, E. Plasma lipoprotein B and VLDL-and LDL-and HDL-cholesterol as risk factors in the development of coronary artery disease in male patients examined by angiography. Atherosclerosis. 39:411 (1981).
4. Debacker, G., Rosseneu, M., and Deslypere, J.P. Discriminative value of lipids and apoproteins in coronary heart disease. Atherosclerosis. 42:197 (1982).
5. Vergani, C., and Bettale, G. Familial hypoalphalipoproteinemia, Clin.Chem.Acta. 114:45 (1981).
6. Third, J.L.H.C., Montag, J., Flynn, M., Freidel, J., Laskarzewski, P., and Gluek, C.J. Primary and familial hypoalphalipoproteinemia. Metabolism. 33:136 (1984).
7. Ordovas, J.M., Schaefer, E.J., Salem, D., Word, R.H., Gleuck, C.J., Vergani, C., Wilson, P.W.F., and Karathanasis, S.K. Apolipoprotein A-I gene polymorphism associated with premature coronary artery disease and familial hypoalphalipoproteinemia. N.Engl.J.Med. 314:671 (1986).
8. Law, S.W., and Brewer, H.B.,Jr. Nucleotide sequence and the encoded aminoacids of human apolipoprotein A-I mRNa, Proc.Natl.Acad.Sci.USA. 8:66 (1984).
9. Karathanasis, S.K. Apolipoprotein multigene family: tandem organization of human apolipoprotein A-I, C-III and A-IV gene. Proc.Natl.Acad.Sci.USA. 82:6374 (1985).
10. Franceschini, G., Sirtori, C.R., Capurso, A., Weisgraber, K.H., and Mahley, R.W. A-I Milano apoprotein. Decreased high density lipoprotein cholesterol levels with significant lipoprotein modifications and without clinical atherosclerosis in an Italian family. J.Clin.Invest. 66:892 (1980).
11. Kessling, A.M., Horsthemke, B., and Humphries, S.E., A study of DNA polymorphism around the human apolipoprotein A-I gene in hyperlipemic and normal individuals. Clin.Genet. 28:296 (1985).
12. Sidoli, A., Guidici, G., Soria, M., and Vergani, C. Restriction fragment lenght polymorphism in the A-I C-III gene complex occurring in a family with hypoalphalipoproteinemia. Atherosclerosis. 62:81 (1985).

13. Brewer, H.B.,Jr., Ronan, R., Meng, M., and Bishop, C. Isolation and characterization of apolipoprotein A-I, A-II and A-IV, in: "Methods in Enzimology", Segrest, J.P. and Albers, J.J. eds. Academic Press, Inc. Orlando, Florida.
14. Sprecher, D.L., Taam, L., and Brewer, H.B., Jr. Two-dimensional electrophoresis of human plasma apolipoproteins. Clin.Chim.Acta. 30:2084 (1984).
15. Schaefer, E.J., Zech, L.A., Jenkins, L.L., Rubalcaba, E.A., Lindgren, F.T., Aamodt, R.L., and Brewer, H.B., Jr. Human apolipoprotein A-I and A-II metabolism. J.Lipid Res. 23:850 (1982).
16. Berman, M., and Weiss, M. SAAM Manual. DHEW Publ. n.(NIH) 78-180. National institutes of Health, Bethesda, Maryland (1978).
17. Weisgraber, K.H., Rall, S.C., Jr., Bersot, T.P., Mahaley, R.W., Franceschini, G., and Sirtori, C.R. Apolipoprotein A-I Milano. Detection of normal A-I in affected subjects and evidence for a cysteine for arginine substitution in the variant A-I. J.Biol.Chem. 258:2508 (1983).
18. Franceschini, G., Sirtori, C.R., Bosisio, E., Gualandri, V., Orsino, G.B., Mogavero, A.M., and Capurso, A. Relation ship of the phenotypic expression of the A-I Milano apoprotein with plasma lipids and lipoprotein patterns. Atherosclerosis. 58:159 (1985).
19. Weisgraber, K.H., Bersot, T.P., Mahaley, R.W., Franceschini, G., Sirtori, C.R. A-I Milano apoprotein. Isolation and characterization of cysteine-containing variant of the A-I apoprotein from human high density lipoproteins. J.Clin.Invest. 66:901 (1980).

SYNTHESIS OF APOLIPOPROTEIN A-I IN THE SKELETAL MUSCLE OF THE DEVELOPING CHICK

S. Calandra and P. Tarugi

Istituto di Patologia Generale

Università di Modena, Via Campi 287, Modena (Italy)

INTRODUCTION

In 1982 Blue et al. reported for the first time that in rooster, apolipoprotein A-I (apo A-I) is synthesized not only by liver and intestine, as in mammals, but also by several peripheral tissues, including the heart and the skeletal muscle (1). They also found that apo A-I produced by peripheral tissues has charge properties and isoform composition superimposable to those of apo A-I synthesized by liver and intestine (1). Since then, other reports have confirmed Blue's observation with regard to the synthesis of apo A-I in the skeletal muscle and have pointed out that a " burst " in apo A-I synthesis in skeletal muscle occurs around the time of hatching. This increased synthesis is associated with a parallel change of the level of translatable apo A-I mRNA (2,3). Apo A-I mRNA level and apo A-I synthesis decrease rapidly after the first week of post-natal life (2, 3). The physiological significance of elevated apo A-I production by the skeletal muscle of the newborn chick is still poorly understood. It has been suggested that the synthesis of apo A-I in skeletal muscle may be important for: a) the local traffic and distribution of lipids among the muscle cells and the cellular organelles of a rapidly growing tissue; b) the removal of lipids accumulated in the skeletal muscle during the prenatal life; c) the differentiation of myoblasts into mature muscle fibers.

Over the last few years we have investigated apo A-I gene expression in several tissues of the adult and the developing chick. In the present review we summarize some observations on the amino acid sequence of chick apo A-I and the possible regulation of apo A-I synthesis in the skeletal muscle of the developing chick.

AMINO ACID SEQUENCE OF CHICK APO A-I

The amino acid sequence of chick apo A-I as derived from
the nucleotide sequence of a full length apo A-I cDNA (4) consists of 264 amino acids (5) as illustrated in Fig. 1.
As found in other species chick apo A-I contains a signal peptide (amino acids 1- 18) which is characterized by a high degree of hydrophobicity and a pro-peptide (amino acids 19-24).
Mature chick apo A-I therefore consists of 240 amino acids
with an estimated molecular weight slightly above 27,000. The
pro-peptide of chick apo A-I differs from the human counterpart in two positions. At position 2 serine replaces histidine and at position 6 histidine replaces glutamine (5, 6). The
difference at position 6 might have an important role in the
cleavage of chick pro-peptide which can be accomplished also
intracellularly (7). As found in humans, the removal of the
pro-peptide leads to a mature peptide which is more acidic
than its precursor (8, 9).
The degree of homology of mature chick apo A-I to the corresponding human and rat sequences is 50% and 45% respectively
(5, 10). Despite this divergence between avian and mammalian
apo A-I most of chick apo A-I (from residue 89 to the end)
can be accomodated in the same periodical structure consisting of multiple repeats of 22 amino acids, each of which is
a tandem array of two 11 amino acid repeats. This repetitive
domain is characterized by the following consensus sequence:
Pro-X-hydrophobic-acid-acid-hydrophobic-base-acid-base-hydro-
phobic-X (5, 10). In addition the alignement of human, chick
and rat apo A-I sequences indicates that the majority of the
amino acid substitutions are conservative (10).

APO A-I mRNA CONCENTRATION IN CHICK TISSUES

Molecular hybridization experiments confirmed that apo
A-I mRNA is detectable in several tissues of the adult chick
(small intestine, liver, brain, heart and skeletal muscle)
(4). In skeletal muscle apo A-I is approximately one tenth
that found in the small intestine, the main apo A-I producing
organ in the adult animal (4). Similar studies conducted in
the developing chick demonstrated that during the embryonic
life chick apo A-I mRNA is found in high concentration only
in the liver whereas its level in the intestine becomes relevant only after the first week of post-natal life (5). In
keeping with previous observations (2, 3) we confirmed that
in skeletal muscle apo A-I mRNA is present at a very high
concentration only around hatching (5). DNase sensitivity
studies of apo A-I gene in isolated nuclei of chick skeletal
muscle indicated that the time-limited rise in skeletal
muscle apo A-I mRNA presumably reflects an increased rate of
transcription (5). Immunocytochemical studies have demonstrated that the synthesis of apo A-I occurs in the muscle cells
and not in other cell types (11).

AMINO ACID SEQUENCE OF CHICK APO A-I

Met-Arg-Gly-Val-Leu-Val-Thr-Leu-Ala-Val-Leu-Phe-Leu-Thr-   14

Gly-Thr-Gln-Ala-Arg-Ser-Phe-Trp-Gln-His-Asp-Glu-Pro-Gln-

Thr-Pro-Leu-Asp-Arg-Ile-Arg-Asp-Met-Val-Asp-Val-Tyr-Leu-   42

Glu-Thr-Val-Lys-Ala-Ser-Gly-Lys-Asp-Ala-Ile-Ala-Gln-Phe-

Glu-Ser-Ser-Ala-Val-Gly-Lys-Gln-Leu-Asp-Leu-Lys-Leu-Ala-   70

Asp-Asn-Leu-Asp-Thr-Leu-Ser-Ala-Ala-Ala-Ala-Lys-Leu-Arg-

Glu-Asp-Met-Ala-(Pro)-Tyr-Tyr-Lys-Glu-Val-Arg-Glu-Met-Trp-   98

Leu-Lys-Asp-Thr-Glu-Ala-Leu-Arg-Ala-Glu-Leu-Thr-Lys-Asp-

Leu-Glu-Glu-Val-Lys-Glu-Lys-Ile-Arg-(Pro)-Phe-Leu-Asp-Gln-   126

Phe-Ser-Ala-Lys-Trp-Thr-Glu-Glu-Leu-Glu-Gln-Tyr-Arg-Gln-

Arg-Leu-Thr-(Pro)-Val-Ala-Gln-Glu-Leu-Lys-Glu-Leu-Thr-Lys-   154

Gln-Lys-Val-Glu-Leu-Met-Gln-Ala-Lys-Leu-Thr-(Pro)-Val-Ala-

Glu-Glu-Ala-Arg-Asp-Arg-Leu-Arg-Gly-His-Val-Glu-Glu-Leu-   182
****
Arg-Lys-Asn-Leu-Ala-(Pro)-Tyr-Ser-Asp-Glu-Leu-Arg-Gln-Lys-
****

Leu-Ser-Gln-Lys-Leu-Glu-Glu-Ile-Arg-Glu-Lys-Gly-Ile-(Pro)-   210

Gln-Ala-Ser-Glu-Tyr-Gln-Ala-Lys-Val-Met-Glu-Gln-Leu-Ser-

Asn-Leu-Arg-Glu-Lys-Met-Thr-(Pro)-Leu-Val-Gln-Glu-Phe-Arg-   238

Glu-Arg-Leu-Thr-Pro-Tyr-Ala-Glu-Asn-Leu-Lys-Asn-Arg-Leu-

Ile-Ser-Phe-Leu-Asp-Glu-Leu-Gln-Lys-Ser-Val-Ala   264

Fig. 1. The amino acid sequence of chick pre-pro-apolipopro-
tein A-I as derived from the nucleotide sequence of a full
length cDNA probe (5). The signal peptide and the propeptide
are underlined. The first amino acid (Pro) of 22 amino acid
repeats is encircled. The sequence of chick apo A-I which
has strong homology to the B,E receptor domain of human apo
E is indicated by stars.

REVERSE CHOLESTEROL TRANSPORT
IN THE DEVELOPING CHICK

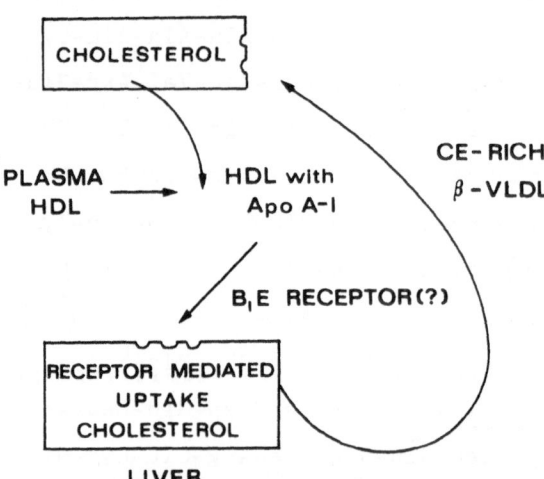

Fig. 2. This simple scheme illustrates the role of apo A-I
containing HDL in removing cholesterol overload from the
skeletal muscle of the newborn chick. HDL with apo A-I synthesized and secreted by the skeletal muscle are taken up by
the liver. It is suggested that chick apo A-I mediates the
hepatic uptake of HDL by interacting with hepatic B,E receptors. The figure also shows that during late embryonic and
early post-natal life cholesteryl ester rich β-VLDL secreted by the liver may be responsible for the cholesterol
overload in tissues.

LIPID OVERLOAD AND APO A-I SYNTHESIS IN THE SKELETAL MUSCLE
OF THE DEVELOPING CHICK.

Since during late embryonic and early post-natal life
there is a massive transfer of lipids from the yolk sac to
the developing chick, the developmentally associated changes
in tissue lipids was studied in various chick tissues including the skeletal muscle. We found that free and esterified
cholesterol accumulate in the skeletal muscle during the embryonic life. Cholesteryl esters concentration decreases
very rapidly 2 days after hatching whereas that of free chole -
sterol decreases more slowly. Interestingly the depletion of
cholesteryl ester store superimposes to the "burst"of apo A-I
synthesis suggesting that these two events may be physiologically related. We think that the accumulation of cholesteryl
esters in skeletal muscle is a major stimulatory factor for
apo A-I synthesis. Teleologically speaking chick skeletal
muscle has developed a mechanism whereby any increase of the
intracellular cholesteryl ester store above a certain thre-
shold level induces the synthesis of an amphipatic protein
(such as apo A-I) capable of binding lipids and vehiculating
them outside the cells. The concept that the synthesis of certain apolipoproteins can be induced by cellular cholesterol
overload is not a novel one. For example it is well established that apo E synthesis and secretion can be induced to
very high levels by loading mouse peritoneal macrophages with
cholesterol (12, 13). The molecular mechanism underlying this
induction is not understood. It is possible that some nucleotide sequence flanking the 5'end of the apo A-I gene in birds
and the apo E gene in mammals functions as regulatory element
specifically sensitive to cholesterol itself or some cholesterol bound trans-acting elements. Elevated apo A-I synthesis
has been observed in skeletal muscles of chick with congeni-
tal muscular dystrophy (14). Interestingly muscular dystrophy
is associated with an accumulation of cholesteryl esters in
the dystrophic muscle (15, 16).

APO A-I CONTAINING LIPOPROTEINS SECRETED BY CHICK SKELETAL
MUSCLE.

Explants of skeletal muscle isolated from newborn chick
and incubated with 35-S methionine secrete 35-S labelled apo
A-I containing HDL (11). Colchicine, a known inhibitor of
protein secretory pathway, abolishes the secretion of HDL by
skeletal muscle. When muscle explants are pre-incubated with
14-C cholesterol to label the intracellular cholesterol pool
14-C cholesterol is secreted into the medium as a component
of HDL particles. Under the electron microscope the d < 1.210
g/ml material secreted by muscle explants is found to contain spheroidal particles resembling plasma HDL. Thus it
seems reasonable to conclude that explants of skeletal muscle
isolated from newborn chick secrete apo A-I molecules which

are part of lipoprotein particles or become loaded with lipid
shortly after their secretion. There is also circumstantial
evidence that apo A-I containing HDL are secreted by the skeletal muscle in vivo. In a time-course study of the changes
of plasma lipoproteins during chick development we observed
that around hatching there is an abrupt and transient increase of HDL and apo A-I in plasma (11). It is likely that this
is due to the secretion of HDL by the skeletal muscles of the
whole organism.

APO A-I IN BIRDS AND APO E IN MAMMALS MAY SHARE SIMILAR
FUNCTIONS.

The expression of chick apo A-I in peripheral tissues
is similar to that of mammalian apo E but not that of mammalian apo A-I (17). This would suggest that chick apo A-I has
functions similar to those of mammalian apo E. This hypothesis is reinforced by the observation that chick plasma lipoproteins lack apo E (or contain negligible amounts of apo E)
and apo E gene is not expressed in chick tissues (11). It was
recently pointed out that one specific sequence of chick apo
A-I (amino acids 170-187) has strong homology to the domain
of human apo E (amino acids 132-149) which mediates the binding of apo E to the B,E receptors (7). Such sequence is
absent in mammalian apo A-I (7). Up to now there is no experimental evidence that chick apo A-I interacts with B,E
receptors in chick tissues nor that it competes with chick
apo B for binding to LDL receptors (18). If indeed chick apo
A-I interacts with B,E receptors possibly present in chick
liver, the secretion of apo A-I containing lipoproteins by
cholesterol overloaded skeletal muscle represents a further
example of an efficient reverse cholesterol transport (Fig.2).

REFERENCES

1. M.L. Blue, P. Ostapchuck, J.S. Gordon, and D.L. Williams,
   Synthesis of apolipoprotein A-I by peripheral tissues of
   the rooster, J. Biol. Chem. 257: 11151 (1982).
2. J.E. Shackelford, and H.G. Lebherz, Regulation of apolipoprotein A-I synthesis in avian muscles, J. Biol. Chem. 258:
   14829 (1983).
3. J.E. Shackelford, and H.G. Lebherz, Synthesis and secretion of apolipoprotein A-I by chick breast muscle, J. Biol.
   Chem. 258: 7175 (1983).
4. St. Ferrari, E. Drusiani, S. Calandra, and P. Tarugi, Isolation of a cDNA clone for chick intestinal apolipoprotein
   A-I (apo A-I) and its use for detecting apo A-I mRNA expression in several chick tissues, Gene 42: 209 (1986).
5. St. Ferrari, P. Tarugi, E. Drusiani, S. Calandra, and M.
   Fregni, The complete sequence of chick apolipoprotein A-I
   mRNA and its expression in the developing chick, Gene 60:
   39 (1987).

6. R.W. Mahely, T.L. Innerarity, C.S. Rall, and K.H.
   Weisgraber, Plasma lipoproteins: apolipoprotein structure and function, J. Lipid Res. 25: 1277 (1984).
7. D. Banerjee, G. Grieninger, J. Lee Parkes, T.K.
   Mukherje, and C.M. Redman, Regulation of apo A-I processing in cultured hepatocytes, J. Biol. Chem. 261: 9844
   (1986).
8. St. Ferrari, P. Tarugi, R. Battini, M. Ghisellini, and
   S. Calandra, Analysis of apolipoprotein A-I synthesized
   in vitro from chick intestinal mRNA. In:" Human Apolipoprotein mutants" C.R. Sirtori, A.V. Nichols and G. Franceschini, eds. Plenum Publishing Corporation, New York
   (1986).
9. S. Calandra, P. Tarugi, St. Ferrari, and M. Ghisellini,
   Isoforms of apolipoprotein A-I in rat and chick plasma
   Atherosclerosis Review 16: 125 (1987).
10. T.B. Rajavashisth, P.A. Dawson, D.L. Williams, J.E.
    Shackelford, H. Lebherz, and A.J. Lusis, Structure,
    evolution, and regulation of chicken apolipoprotein A-I,
    J. Biol. Chem. 262: 7058 (1987).
11. P. Tarugi, D. Reggiani, and S. Calandra, Transient hyper-
    lipoproteinemia and tissue cholesterol accumulation in
    the developing chick. A possible role of apo A-I (abr.)
    Symposium on " Hyperlipoproteinemia and Atherosclerosis
    Cambridge, 8-10 December (1987).
12. S.K. Basu, M.S. Brown, Y.K. Ho, R.J. Havel, and J.L.
    Goldstein, Mouse macrophages synthesize and secrete a
    protein resembeling apolipoprotein E, Proc. Natl. Acad.
    Sci. (USA) 78: 7545 (1981).
13. S.K. Basu, Y.K. Ho, M. S. Brown, D.W. Bilheimer, R.G.W.
    Anderson, and J.L. Goldstein, Biochemical and genetic
    studies of the apoprotein E secreted by mouse macropha-
    ges and human monocytes, J. Biol. Chem. 257: 9788 (1982).
14. J.E. Shackelford, and H.G. Lebherz, Synthesis of apoli-
    poprotein A-I in skeletal muscles of normal and dystro-
    phic chickens, J. Biol. Chem. 260: 288 (1985).
15. G.E. Sumnicht, and R.A. Sabbadini, Lipid composition of
    transverse tubular membranes from normal and dystrophic
    skeletal muscle, Arch. Biochem. Biophys. 215: 628 (1982).
16. D.E. Kuhn, and D.M. Logan, Fiber-specific cholesterol
    changes in murine dystrophy, Biochim. Biophys. Acta
    921: 13 (1987).
17. N.A. Elshourbagy, W.S. Liao, R.W. Mahley, and J.M.
    Taylor, Apolipoprotein E mRNA is abundant in the brain
    and adrenals, as well as in the liver, and is present in
    other peripheral tissues of rats and marmoset, Proc.
    Natl. Acad. Sci. (USA) 82: 203 (1985).
18. G. Rajan, D.L. Barber, and W.J. Schneider, Characteriza-
    tion of the chicken oocyte receptor for low and very low
    density lipoproteins, J. Biol. Chem. 262: 16838 (1987).

APOLIPOPROTEIN-SPECIFIC HIGH DENSITY LIPOPROTEIN POPULATIONS IN PLASMA OF CARRIERS OF THE APOLIPOPROTEIN AI-MILANO

Alex V. Nichols[a], Marian C. Cheung[b], Patricia J. Blanche[a],
Elaine L. Gong[a], Guido Franceschini[c], and Cesare R. Sirtori[c]

[a] Donner Laboratory, Lawrence Berkeley Laboratory, University of
California, Berkeley, CA (USA); [b] Department of Medicine, School of
Medicine, University of Washington, Seattle, WA (USA); and
[c] Center E. Grossi Paoletti, University of Milan, Milan (Italy)

Introduction

Characteristic features of the apolipoprotein AI-Milano ($apoAI_M$) high density lipoproteins (HDL) are: (i) their low plasma concentration; (ii) their particle size profile that consists primarily of two major peaks, one in the HDL_{3a} size interval (8.8-8.2 nm) and the other in the HDL_{3b} size interval (8.2-7.8 nm); and (iii) their unique protein moiety that includes the $apoAI_M$ variant, identified by an Arg 173 → Cys substitution (1,2). While the presence of a low plasma HDL concentration and a particle size distribution enriched in smaller particles is frequently observed in hypertriglyceridemic subjects (3,4), these features are observed in $apoAI_M$-carriers even at normal triglyceride levels (Figure 1). Thus, additional factor(s) appear to be involved in determining the characteristics of the $apoAI_M$-HDL particles. One factor may be an effect of the variant $apoAI_M$ on the structure, apolipoprotein composition and metabolism of the $apoAI_M$-HDL.

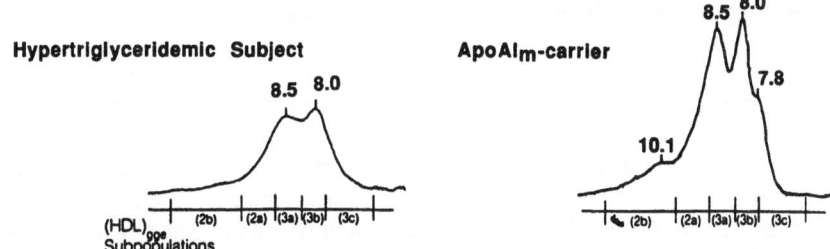

Figure 1. Particle size profiles of HDL, from an $apoAI_M$-carrier (patient VII-134; triglyceride, 107 mg/dl; HDL-cholesterol, 14 mg/dl) and a hypertriglyceridemic subject (triglyceride, 358 mg/dl; HDL-cholesterol, 33 mg/dl); profiles determined by gradient gel electrophoresis (4-30% polyacrylamide; stained for protein). Profiles have not been adjusted to reflect plasma concentration and are only presented to show relative contributions of the different HDL subpopulations. Particle size intervals of major HDL subpopulations (HDL_{2b}, HDL_{2a}, HDL_{3a}, HDL_{3b}, and HDL_{3c}) are indicated below profiles. Particle sizes (in nm) of major components in HDL size profiles are indicated at peak maxima in this and subsequent figures.

In normal subjects, the bulk of the HDL particle size distribution is comprised of two major apolipoprotein-specific populations, HDL(AI w/o AII), HDL particles containing apoAI without apoAII, and HDL(AI w AII), HDL particles containing apoAI with apoAII. These populations can be isolated by immuno-affinity chromatography (5) and analyzed for their particle size distribution by gradient gel electrophoresis (6). HDL(AI w/o AII) of normal subjects (Figure 2, left) usually exhibit profiles with two major peaks, one (mean size 10.4 nm) in the HDL_{2b} interval and the other (mean size 8.5 nm) in the HDL_{3a} interval. Analysis of the protein moieties of these two HDL subpopulations indicates a molecular weight equivalence predominantly of four and three molecules of apoAI per particle, respectively (7). HDL(AI w AII) profiles of normal subjects (Figure 2, right) generally show three peaks with corresponding mean sizes of 9.6 nm (in HDL_{2a} interval), 8.9 nm (in HDL_{2a} interval) and 8.0 nm (in HDL_{3b} interval). Since the presence of $apoAI_M$-$apoAI_M$ homodimers and $apoAI_M$-$apoAII_s$ heterodimers ($apoAI_M$ with disulfide bond to $apoAII_s$, the single 77 amino acid monomer of dimeric apoAII) on $apoAI_M$-HDL particles might influence not only the stoichiometry of their associated apolipoproteins but also the physical properties of these particles, we investigated the characteristics of the two apolipoprotein-specific populations in plasma of two normotriglyceridemic $apoAI_M$-carriers after isolation by immunoaffinity chromatography (7).

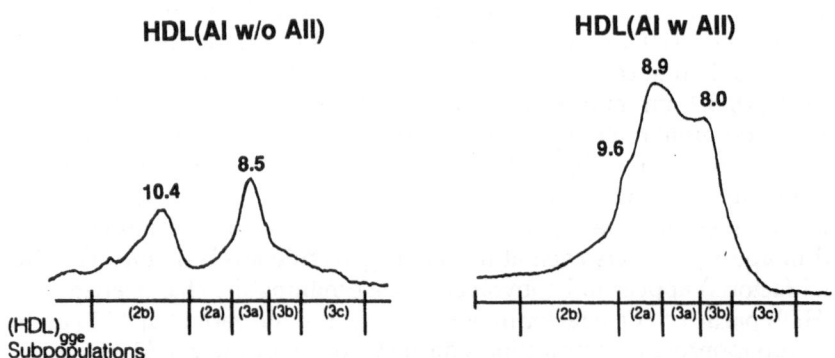

Figure 2. Representative particle size profiles of the two major apolipoprotein-specific HDL populations (HDL(AI w/o AII) and HDL(AI w AII)) isolated by immunoaffinity chromatography from plasma of a normolipidemic subject.

Plasma Lipids and Lipoproteins of Carriers and Normal Subjects

Table 1 shows the lipid and lipoprotein concentrations in plasma of $apoAI_M$-carriers and normal subjects investigated in the present study. The low plasma HDL-cholesterol (HDL-C) levels in the $apoAI_M$-carriers are reflected in reductions in associated HDL apolipoproteins, apoAI and apoAII. The distribution of apoAI between the HDL(AI w/o AII) and the HDL(AI w AII) populations as isolated by immunoaffinity chromatography is about equal and similar to that observed in the normal subjects. Our ability to isolate the two populations in carrier HDL by immunoaffinity chromatography is consistent with previous observations that $apoAI_M$ is immunochemically identical to normal apoAI (2).

TABLE 1. Plasma Lipid and Protein Concentration in ApoAI_M-Carriers and Normal Subjects

	mg/dl				
	C	TG	HDL-C	AI	AII
ApoAI_M-Carrier					
VII-48	139	61	17	67	18
VII-134	155	107	14	75	18
Normal					
N-1	182	93	60	186	47
N-2	148	36	53	134	25
N-3	179	32	90	208	31

Abbreviations: C, total cholesterol; TG, triglyceride; HDL-C, HDL cholesterol; AI, apoAI; AII, apoAII.

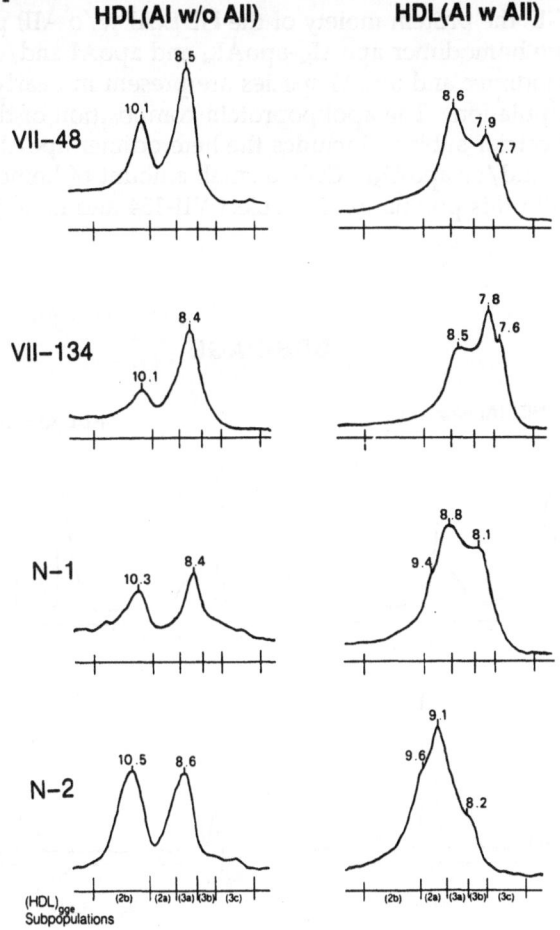

Figure 3. Particle size profiles of the HDL(AI w/o AII) and HDL(AI w AII) populations isolated from plasma of two apoAI_M-carriers (subjects VII-48 and VII-134) and two normal subjects (N-1 and N-2) by immunoaffinity chromatography.

Particle Size Profiles of HDL(AI w/o AII) and HDL(AI w AII) Populations

Figure 3 shows profiles of HDL(AI w/o AII) and HDL(AI w AII) populations from two apoAI$_M$-carriers and two normal subjects. The HDL(AI w/o AII) population in carriers shows two major peaks, one in the HDL$_{2b}$ and the other in the HDL$_{3a}$ interval; the particle sizes of the HDL$_{2b}$ are smaller than those of the normal subjects, while the sizes of the HDL$_{3a}$ components are similar to normal subjects. The relative contribution of the larger HDL$_{2b}$(AI w/o AII) particles to the overall population profile is lower in the carriers. The HDL(AI w AII) profiles of the carriers show the presence of three components whose peak areas fall within the particle size range of the HDL$_{2a}$, HDL$_{3a}$ and HDL$_{3b}$ size intervals. The particle sizes of the carrier HDL(AI w AII) components are smaller than those of the three components observed in profiles of the normal subjects. A prominent feature of the carrier HDL(AI w AII) population is the presence of a distinct peak (mean size 7.8-7.9 nm) in the HDL$_{3b}$ size interval.

Apolipoproteins in HDL(AI w/o AII) and HDL(AI w AII) Populations

By SDS-PAGE, the protein moiety of the HDL(AI w/o AII) population of carriers includes the homodimer apoAI$_M$-apoAI$_M$ and apoAI and/or apoAI$_M$ (Figure 4); the homodimer and apoAI species are present in nearly equivalent amounts in this population. The apolipoprotein composition of the HDL(AI w AII) population of carrier subjects includes the heterodimer apoAI$_M$-apoAII$_s$, apoAII, and apoAI and/or apoAI$_M$. Only a small amount of homodimer apoAI$_M$-apoAI$_M$ is detected in this population in subject VII-134 and none in subject VII-48.

SDS-PAGE

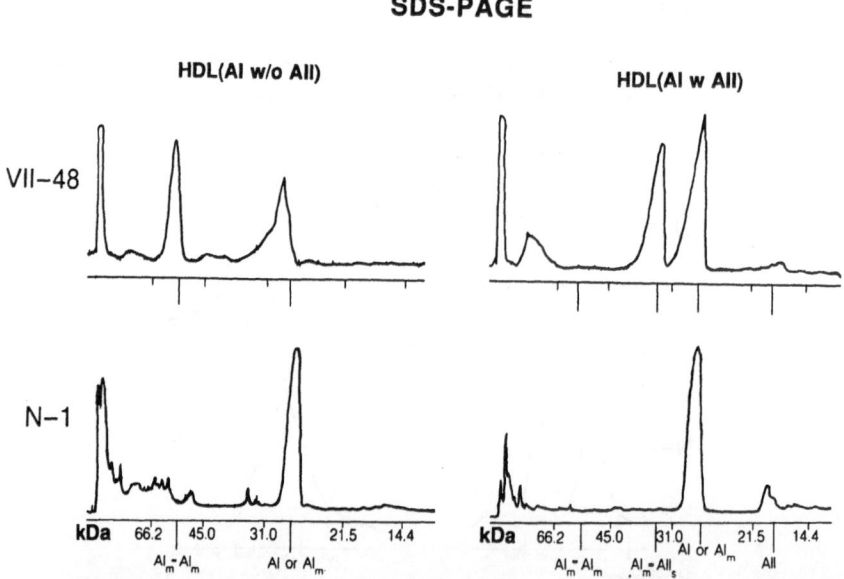

Figure 4. SDS-polyacrylamide gel electrophoresis profiles (using 15% acrylamide and 0.035% Coomassie R250) of protein moieties of HDL(AI w/o AII) and HDL(AI w AII) populations isolated from plasma of apoAI$_M$-carrier (VII-48) and normolipidemic subject (N-1). Molecular weights (in kDA) of major apolipoprotein constituents are indicated below profiles.

Subpopulations of the apolipoprotein-specific populations from carriers and normal subjects were isolated by gel filtration chromatography and analyzed by chemical crosslinking (8) to assess apolipoprotein content per particle (Figure 5). In carriers, the HDL$_{2b}$(AI w/o AII) subpopulation shows the presence predominantly of particles with protein molecular weight equivalent to four molecules of apoAI. The protein moiety of the HDL$_{3a}$(AI w/o AII) subpopulation, in large part, exhibits a molecular weight equivalent to three molecules of apoAI per particle. Less abundant small particles, in the HDL(AI w/o AII) population, that migrate within the HDL$_{3c}$ interval (7.8-7.2 nm), mainly show a protein molecular weight equivalent to 2 molecules of apoAI (data not shown). In normal subjects, corresponding subpopulations of the HDL(AI w/o AII) population show similar results for equivalent molecular weights of their protein moieties. Thus, notwithstanding the presence of the apoAI$_M$-apoAI$_M$ homodimer in the HDL(AI w/o AII) population, the apolipoprotein stoichiometry of the major subpopulations in carriers is similar to that for the same subpopulations in normal subjects.

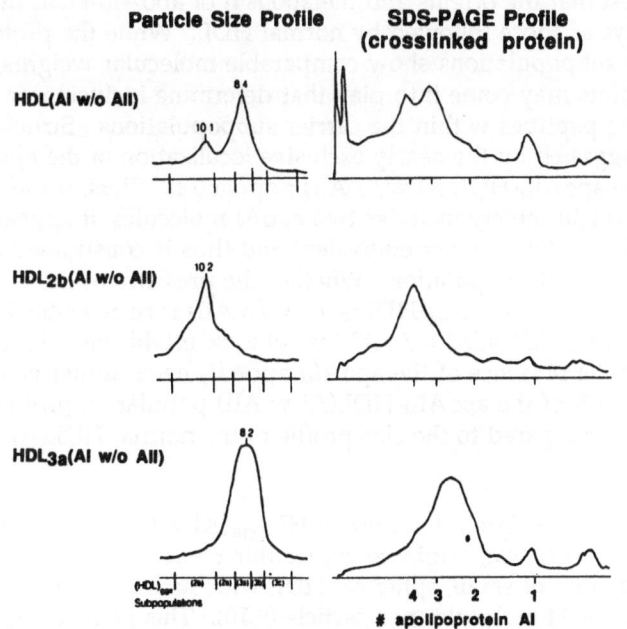

Figure 5. Particle size profiles (at left) of the HDL(AI w/o AII) population and its two major subpopulations (HDL$_{2b}$(AI w/o AII) and HDL$_{3a}$(AI w/o AII)) isolated by gel filtration from the HDL(AI w/o AII) population of apoAI$_M$-carrier VII-134. SDS-polyacrylamide gel electrophoresis profiles (at right) are of crosslinked protein moieties obtained from the above HDL(AI w/o AII) population and its subpopulations. Reference scale for number of apoAI per particle is based on migration positions of crosslinked oligomers of apoAI.

Gel filtration chromatography of carrier HDL(AI w AII) population yielded fractions enriched in specific subpopulations but containing contributions from adjacent subpopulations. The protein molecular weight for the crosslinked fractions ranged from 82 to 86 kDa. A detailed stoichiometry for the apolipoprotein composition for these particles was not available and hence a definitive interpretation of the crosslink data is not possible. However, the molecular weight results would be consistent with a protein moiety comprised of one molecule of apoAI or apoAI$_M$, one molecule of apoAI$_M$-apoAII$_s$ and one molecule of apoAII.

As noted earlier, the contribution of the apoAI$_M$-apoAI$_M$ homodimer to the HDL(AI w AII) population is very low and hence it is not included in the above hypothetical apolipoprotein composition. For the major HDL(AI w AII) subpopulations from normal subjects, the crosslink data show a protein molecular weight (mean value of 84 kDa) that would be consistent with the presence of a mixture of particles with molar ratios of apoAI:apoAII of 2:1 and 1:1. Thus, as in the case of carrier HDL(AI w/o AII) subpopulations, subpopulations of HDL(AI w AII) exhibit protein crosslink properties similar to those of normal subjects.

Discussion

ApoAI$_M$-HDL populations, which include three unusual peptide constituents (apoAI$_M$, apoAI$_M$-apoAI$_M$, and apoAI$_M$-apoAII$_s$) in their protein moiety, exhibit particle size profiles and apparent apolipoprotein stoichiometry that in many but not all respects resemble those of normal HDL populations (7). Such similarities suggest that the origins and metabolism of apoAI$_M$-HDL may follow the same pathways as those followed by normal HDL. While the protein moieties of corresponding subpopulations show comparable molecular weights, certain structural constraints may come into play that determine inclusion or exclusion of apoAI$_M$-containing peptides within the carrier subpopulations. Structural constraints are suggested by the nearly exclusive localization of the apoAI$_M$-apoAI$_M$ homodimer in the apoAI$_M$-HDL(AI w/o AII) population. Thus, if the apoAI$_M$-HDL(AI w AII) protein moiety includes two apoAI molecules, it appears that the apoAI$_M$-apoAI$_M$ homodimer is not equivalent and thus is constrained to the apoAI$_M$-HDL(AI w/o AII) population. Whether the presence of the apoAI$_M$-apoAI$_M$ homodimer in the apoAI$_M$-HDL$_{2b}$(AI w/o AII) is responsible for its smaller size relative to normal HDL$_{2b}$(AI w/o AII) is yet to be established. Likewise, it is not clear whether the presence of the apoAI$_M$-apoAII$_s$ heterodimer is a major factor in the significant shift of the apoAI$_M$-HDL(AI w AII) population profile to smaller particle size, when compared to the size profile of the normal HDL(AI w AII) population.

It is generally considered that normal HDL$_{3a}$(AI w/o AII) species, with three apoAI molecules per particle, originate via lecithin:cholesterol acyltransferase-induced transformation of small, spherical HDL with sizes in the range of 7.6-7.2 nm and with two apoAI molecules per particle (9,10). This pathway appears to involve a fusion step wherein a product with three apoAI molecules is formed by merger of precursors each with two apoAI molecules and release of one apoAI molecule. Since apoAI$_M$-apoAI$_M$ homodimers are present in the apoAI$_M$-HDL$_{3a}$(AI w/o AII) subpopulation, it is possible that in carriers small, spherical precursor HDL particles may also include species with only the homodimer. However, for this precursor to proceed via this pathway to a HDL$_{3a}$(AI w/o AII) product, the presence of small spherical particles with two uncomplexed apoAI molecules would also be required.

A further interesting question also arises with respect to potential changes in function of apoAI$_M$-HDL(AI w AII) species that might result from inclusion of the the heterodimer apoAI$_M$-apoAII$_s$. It is possible that metabolic processes requiring intact apoAII on HDL(AI w AII) particles may now be attenuated by the presence of the heterodimer.

Thus, as a consequence of the unique properties of the apoAI$_M$ variant, many fascinating questions concerning HDL structure and function as well as questions concerning the putative protective role of plasma HDL in cardiovascular disease can be posed.

References

1. G. Franceschini, C.R. Sirtori, A. Capurso, K.H. Weisgraber, and R.W. Mahley, A-I$_{Milano}$ apoprotein: decreased high density lipoprotein cholesterol levels with significant lipoprotein modifications and without clinical atherosclerosis in an Italian family, J. Clin. Invest. 66: 892 (1980).

2. K.H. Weisgraber, T.P. Bersot, R.W. Mahley, G. Franceschini, and C.R. Sirtori, A-I$_{Milano}$ apoprotein: isolation and characterization of a cysteine-containing variant of the A-I apoprotein from human high density lipoproteins, J. Clin. Invest. 66: 901 (1980).

3. A.V. Nichols, P.J. Blanche, R.M. Krauss, and E.L. Gong, Gradient gel electrophoresis of HDL with the ultracentrifugal d < 1.200 fraction from human serum, in "Report of the High Density Lipoprotein Methodology Workshop 1979", K. Lippel, ed., NIH Publication No. 79-1661, 303-309, US Department of Health, Education and Welfare, NIH, Bethesda, MD (1979).

4. L.B.F. Chang, G.J. Hopkins, and P.J. Barter, Particle size distribution of high density lipoproteins as a function of plasma triglyceride concentration in human subjects, Atherosclerosis 56: 61 (1985).

5. M.C. Cheung, and J.J. Albers, Characterization of lipoprotein particles isolated by immunoaffinity chromatography; particles containing A-I and A-II and particles containing A-I but no A-II, J. Biol. Chem. 259: 12201 (1984).

6. A.V. Nichols, P.J. Blanche, and E.L. Gong, Gradient gel electrophoresis of human plasma high density lipoproteins, in: "Handbook of Electrophoresis, Vol. III", L.A. Lewis, ed., CRC Press, Boca Raton, Florida (1983).

7. M.C. Cheung, A.V. Nichols, P.J. Blanche. E.L. Gong, G. Franceschini, and C.R. Sirtori, Characterization of A-I-containing lipoproteins in subjects with A-I Milano variant, Biochim. Biophys. Acta 960: 73 (1988).

8. J.B. Swaney, and K. O'Brien, Cross-linking studies of the self-association properties of apoA-I and apoA-II from human high density lipoprotein, J. Biol. Chem. 253: 7069 (1978).

9. C. Chen, K Applegate, W.C. King, J.A. Glomset, K.R. Norum, and E. Gjone, A study of the small spherical high density lipoproteins of patients afflicted with familial lecithin:cholesterol acyltransferase deficiency, J. Lipid Res. 25: 269 (1984).

10. A.V Nichols, P.J. Blanche, E.L. Gong, V.G. Shore, and T.M. Forte, Molecular pathways in the transformation of model discoidal lipoprotein complexes induced by lecithin:cholesterol acyltransferase, Biochim. Biophys. Acta 834: 285 (1985).

APO B GENE VARIANTS ARE INVOLVED IN DETERMINING SERUM CHOLESTEROL

LEVELS: TOWARDS IDENTIFYING THESE VARIANTS

Philippa Talmud, Richard Houlston, Alison Dunning and
Steve Humphries

Charing Cross Sunley Research Centre, Lurgan Avenue
London, W6 8LW

For any individual in the population the determination of lipid
levels is multifactorial. By studying the population as a whole it is
possible to get a better definition of these factors. In the case of
cholesterol, for example, we know from a number of studies that 50% of
the total phenotypic variance in the population is due to environmental
factors and 50% due to genetic factors, with variation at a number of
genetic loci contributing to this genetic component of cholesterol
variance [1,2]. Apo B plays a central role in lipid metabolism as ligand
for the LDL receptor and epidemiological studies have shown that
increased apo B levels correlate with increased risk of coronary heart
disease [3]. Thus variation in the apo B gene could be involved in
determining cholesterol levels and hence atherosclerosis risk.

We have used molecular genetic techniques to ask three questions.
Firstly to try and establish, with the use of Restriction Fragment
Length Polymorphisms (RFLPs) of the apo B gene, whether there is an
association between apo B gene variants and different types of hyper-
lipidemias; secondly whether there is an association with apo B variants
and lipid levels in the healthy population and finally to try and identify
at the DNA level what mutations are giving rise to these variants.

In an attempt to answer the first question, we compared the relative
allele frequencies of a number of RFLPs of the apo B gene in a sample of
normolipidemic individuals with groups of clinically defined hyperlipidemic
patients. Only with the XbaI RFLP did we see any significant difference
in allele frequency when compared to our control sample (Table 1). This
was in the group of patients with type III hyperlipidemia. Our results
show a significantly higher relative frequency in this group of the X2
allele of the XbaI RFLP than in the normolipidemic individuals or other
patient groups. This suggests that variation at the apo B locus may
be involved in this disorder [4].

This analysis was extended by a more detailed examination of the
group of normolipidemic individuals. Figure 1 shows the mean serum
cholesterol levels of individuals with the three different genotypes for

the XbaI polymorphism. Individuals homozygous for the X2 allele had a
significantly higher mean serum cholesterol level than individuals
homozygous for the X1 allele. Individuals with the genotype X1X2 had an
intermediate mean serum cholesterol level. We estimate that in our sample
the effect of the X1 allele is to lower serum cholesterol levels by
0.14 mmol/l (5.2 mg/dl) whereas the effect of the X2 allele is to raise
cholesterol levels by the same amount[4]. In this normolipidemic population
our data suggests that variation in the gene for apo B, associated with
the XbaI genotype, is involved in the determination of serum cholesterol
levels.

Table 1. Comparison of genotype distribution and relative allele
frequency for the XbaI RFLP in normolipidaemic individuals
and in individuals with different types of hyperlipidemia

	Genotype			Allele Frequency		χ^2 of allele frequencies
	X1X1	X1X2	X2X2	X1	X2	
Normal n=62	12	38	12	0.50	0.50	
IIa n=44	12	24	8	0.54	0.45	$\chi^2=0.422$ 0.5<p<0.2
IIb n=16	5	5	6	0.47	0.53	$\chi^2=0.095$ 0.7<p<0.9
III n=64	8	25	31	0.32	0.68	$\chi^2=8.63$ p<0.01
IV n=17	7	7	3	0.16	0.39	$\chi^2=1.47$ 0.2<p<0.5
V n=12	1	1	3	0.42	0.58	$\chi^2=0.56$ 0.2<p<0.5

χ^2 analysis of "2 x 2" tables based on gene counting

Using data from our control population it is possible to estimate that
genetic variability, defined by and associated with the XbaI RFLP, accounts
for between 4-10% of the total phenotypic variance in serum cholesterol
levels[4]. Recently it has been shown that variation at the apo E gene
locus contributes 4-8% of this phenotypic variation in cholesterol
levels[5,6].

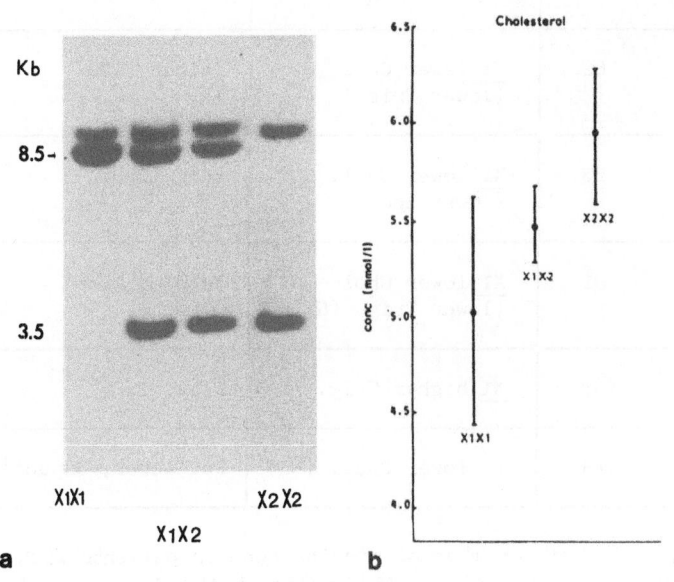

Figure 1. (a) Southern blot analysis of the apo B DNA polymorphism.
The hybridisation pattern obtained from a digest of 5 μg of DNA from four
unrelated individuals is shown. (b) Lipid levels in normolipidemic
individuals with different XbaI genotypes. The mean serum cholesterol
levels ± 95% confidence limits are shown for individuals with the genotype X1X1, X1X2 and X2X2. Cholesterol levels were corrected for age and
sex. In this sample, mean cholesterol levels in normal men and women
were the same at age 48, so all levels were corrected to this age. A
one way analysis of variance was performed on the adjusted lipid levels.
The respective means and standard deviations of cholesterol levels mmol/l
for the three genotypes were 5.03 (± 0.945) 5.48 (± 0.65), and 5.96 (± 0.52).
One-way pairwise analysis of variance results were as follows:

	F-ratio	df	P
X1X1 vs X1X2	3.83	1 : 47	<0.1
X1X1 vs X2X2	3.91	1 : 22	<0.001
X1X2 vs X2X2	4.92	1 : 47	<0.05

A number of studies concerning the association between the XbaI
genotype of the apo B gene and cholesterol levels in the clinically
well population have been reported. Of the 6 studies listed in Table 2
5 report that individuals with the genotype X1X1 have the lowest serum
cholesterol levels in the study population[4,7-11].

Table 2. Reported Associations between apo B XbaI variable site and
lipid levels in controls
(X1 - 8.6 kb absence of cutting site)

ORIGIN	NO	COMMENTS	REFERENCES
London	81	X1 lower Chol. (lower Trig.)	Law 1986[7]
London	62	X1 lower Chol. (lower Trig.)	Talmud, 1987[4]
Norway	56	X1 lower Chol. (lower apo B)	Berg, 1987[8]
Boston	81	X1 lower Chol. (NS) (lower Trig. (NS)	Hegele, 1986[9]
Seattle	100	X1 higher Trig.	Deeb, 1986[10]
Austria	118	X1 lower Chol. (NS)	Paulweber, Unpub[11]

By contrast out of a number of studies done on patients with CAD
only in 1 study has a significant difference in the frequency of the
XbaI RFLP allele been reported when compared to the control population[9-12].
Thus variants of the apo B gene associated with the XbaI genotype
contribute to cholesterol levels within the healthy population but may
not make a contribution to the development of CAD.

It is unlikely that the mutation causing raised cholesterol levels
is at the XbaI polymorphic site itself. We know from the sequence data
that the loss of the XbaI polymorphic site is due to a T to C substitution.
This is a silent mutation that does not alter the amino acid (threonine)
residue 2488[13]. We propose that what we are observing is linkage disequilibrium between the XbaI polymorphic site and a mutation in the apo B
gene giving rise to raised cholesterol levels.

The promoter region and the putative receptor binding domain of the
apo B gene represent the two functionally important regions where mutations
could affect apo B synthesis or binding and thus directly affect cholesterol
levels. We are at present concentrating on these two regions to try and
identify mutations that might appear frequently in the population and
represent common variants of apo B.

A defect in the receptor binding domain of the apo B gene could
result in under removal of apo B. We studied a group of 22 normolipidemic
males all of whom had fractional catabolic rates (FCR) of LDL estimations
using ^{125}I-labelled LDL and the urine plasma ratio method[14] to see if there
was an association between XbaI genotype and FCR. The FCR of LDL gives a
measure of LDL turnover. Individuals with the genotype X1X1 had a
significantly higher FCR than individuals with the genotype X2X2. This
suggests that individuals with the genotype X2X2 have the lowest removal
rate and therefore the highest cholesterol concentrations and this is
probably due to a defect in the receptor binding domain.

The next stage in the analysis will be to determine the sequence of this part of the apo B gene in the individuals studied. A rapid method to do this is the Polymerase Chain Reaction (PCR)[15]. We have used the PCR to amplify an 800 base-pair region spanning the proposed receptor binding domain (amino acids 3130-3396)[16]. The amplified DNA can be subcloned into M13 and sequenced to see if there are any common changes that are common in individuals with different XbaI genotypes.

In conclusion, we have confirmed that variants of the apo B gene identified by the XbaI RFLP are contributing to the total phenotypic variance of cholesterol in the population. The studies on individuals where the FCR has been estimated show that variation within the amino acid sequence of the apo B gene may be making a significant contribution to this variability in the population cholesterol levels. Molecular biology techniques should allow the elucidation of the mutations in the apo B gene that are causing this.

ACKNOWLEDGEMENTS

This work was supported by the Charing Cross Sunley Research Centre and grants from the British Heart Foundation (RG5).

REFERENCES

1. Sing CF & Orr JD. Analysis of genetic and environmental sources of variation in serum cholesterol. Techumseh, Michigan IV. Separation of polygene from common environmental effects. Am.J.Hum.Genet. 30: 491-504.(1978).
2. Moll PP, Powsner R and Sing CF. Analysis of genetic and environmental sources of variation in serum cholesterol in Techumseh, Michigan V. Variance components estimated from pedigrees. Am.J.Hum.Genet. 42: 343-354 (1979).
3. Whayne TF, Alaupovic P, Curry MD, Lee ET, Anderson PS and Schechter E. Plasma apolipoprotein B and VLDL-, LDL- and HDL-cholesterol as risk factor in the development of coronary arterial disease in male patients examined by angiograph. Atherosclerosis 39: 411-424 (1981).
4. Talmud PJ, Barni N, Kessling AM, Carlsson P et al. Apolipoprotein B gene variants are involved in the determination of serum cholesterol levels; a study in normo and hyperlipidaemic individuals. Atherosclerosis 67: 81-89 (1987).
5. Sing CF, Davignon J. Role of apolipoprotein E polymorphism in determining normal plasma lipid and lipoprotein variation. Am.J.Hum.Genet. 37: 268-285 (1985).
6. Boerwinkle E and Utermann G. Simultaneous effects of the apolipoprotein E polymorphism on apolipoprotein E, apolipoprotein B and cholesterol metabolism. Am.J.Hum.Genet. 42: 104-112 (1988).
7. Law A, Powell LM, Wallis SC et al. Common DNA polymorphism within the coding sequence of apolipoprotein B gene associated with altered lipid levels. Lancet, i: 1301 (1986)
8. Berg K. DNA polymorphism at the apolipoprotein B locus is associated with lipoprotein level. Clin.Genet. 30: 515 (1986).
9. Hegele RA, Huang LS, Herbert PN, Blum CB, Buring JE, Hennekens CH and Breslow JL. Apolipoprotein B gene DNA polymorphisms associated with myocardial infarction. New Engl.J.Med. 315: 1509 (1986).
10. Deeb S, Failor A, Brown BG, Brunzell JD et al. Molecular genetics of apolipoproteins and coronary heart disease. Cold Spring Harbor Symposia on Quant. Biol. LI: 403-409.

11. Paulweber B, Friedl W, Drempler F, Humphries SE and Sandhofer F. Association of DNA polymorphism at the apo B gene locus with coronary heart disease and serum VLDL levels. <u>Atherosclerosis</u> submitted.

12. Gallagher J, Myant N, Wile D, Trayner I and Humphries SE. Apo B polymorphism in relation to CAD. In preparation (1988).

13. Carlsson P, Darnfors C, Olofsson SO and Bjursell G. Analysis of human apolipoprotein B gene: complete structure of the B-74 region <u>Gene</u> 49: 29 (1986).

14. Houlston RS, Turner PR, Revill F, Lewis B and Humphries SE The fractional catabolic rate of low density lipoprotein in normal individuals is influenced by variation in the apolipoprotein B gene: a preliminary study. <u>Atherosclerosis</u> 71: 81-85 (1988).

15. Saiki RK, Scharf S, Falosna F, Mullis KB et al Enzymatic amplification of β-globin genomic sequences and restriction analysis for diagnosis of sickle cell anemia. <u>Science</u> 230: 1350-1354 (1985).

16. Knott TJ, Pease RJ, Powell LM, Wallis SC, Rall SC, Innerarity TL, Blackhart B, Taylor WR, Lusis AJ, McCarthy BJ, Mahley RW, Levy-Wilson B and Scott J. Human apolipoprotein B : complete cDNA sequence and identification of structural domains of the protein <u>Nature</u> 323: 734 (1986).

GENETIC EVIDENCE THAT THE APOLIPOPROTEIN GENE IS NOT INVOLVED IN

ABETALIPOPROTEINEMIA

Philippa J. Talmud, June Lloyd[+], David Muller[+], David R. Collins*, James Scott* and Steve Humphries

Charing Cross Sunley Research Centre, London, W6 8LW
[+]Institute of Child Health, London, WC1N 1EH
*MRC Clinical Research Centre, Middlesex, HA1 3UJ

Elevated serum levels of apo B-containing lipoproteins are associated with risk of Ischaemic Heart Disease[1], therefore factors controlling apo B synthesis in the liver and intestine are important to our understanding the development of hyperlipidaemia and atherosclerosis. It may be possible to obtain a better idea of the control of apo B synthesis by studying patients with defects in the synthesis and secretion of apo B containing lipoproteins.

Abetalipoproteinemia (ABL) is a recessive disorder characterised by undetectable amounts of apo B-containing lipoproteins[2]. The clinical features associated with ABL are acanthocytosis present from birth, retinopathy and neuropathy which develop in the second decade and fat malabsorption leading to fat soluble vitamin deficiencies[3]. Obligate heterozygotes have normal apo B and lipoprotein levels and are phenotypically normal.

Our study involved 10 ABL patients and their parents. We have used molecular biological techniques to study the gross structure of the apo B gene, the allele frequency of the apo B gene Restriction Fragment Length Polymorphisms (RFLPs) and to perform linkage analysis in two of these families each with two affected children. All the patients we studied had the classical features of ABL. The parents were non-cosanguinous and had apo B levels within the normal range. We hybridised Southern blots of restriction enzyme digested DNA from the patients with DNA probes that span the 3' and 5' ends of the apo B gene, to see if we could detect any gross deletions or insertions of the gene. In all cases fragment sizes were normal. Within the limit of the probes available we could detect no gross alteration in apo B gene structure in these patients. We next determined the allele frequency of the XbaI, EcoRI and PvuII RFLPs of the apo B gene in both patients and parents and compared them to a control group (Table 1). There was a significant difference in the allele frequency of the PvuII RFLP in the patient group when compared to the controls. We feel that this is probably a chance observation due to small sample size and would not be confirmed in a larger patients sample.

Table 1. Relative allele frequencies of the apo B RFLPs in controls, patients and parents

	XbaI		EcoRI		PvuII	
	X1	X2	R1	R2	P1	P2
Normal (n=100)	.49	.51	.86	.14	.92	.08
Patient* (n=8)	.56	.44	.84	.13	.67 (χ^2=10.2 P<0.01)	.33
Parents (n=14)	.59	.41	.81	.19	.75 (χ^2=3.05 N.S.)	.25

* S.J. and C.M. excluded.

Amongst our patients we had two families each with two affected children. Using the apo B RFLPs we looked at the segregation of the apo B gene in these families. In a recessive disorder affected children inherit the same defective gene from both mother and father. Co-segregation of the disorder and a gene in a typical small nuclear family is illustrated in Figure 1. Results such as this would not be conclusive, as this pattern of inheritance could have occurred by chance alone and a large number of families would have to be studied in order to obtain a lod score of 3.

The lipoprotein and apolipoprotein levels for members of these two families used in the linkage analysis are given in Table 2.

Table 2.   Lipid and Apolipoprotein levels in the two families studied

	Cholesterol mmol/1	HDL Chol mmol/1	Triglyceride mmol/1	Apo B mg/dl	Vitamin E µmol/1
Mr J	6.6	1.2	1.4	112	23.6
Mrs J	4.8	1.0	0.6	74	24.3
MJ	0.9	0.8	0.1	0	1.5*
SJ	0.6	0.4	0.1	0	0.6*
Mr M	4.0	0.9	2.0	84	25.1
Mrs M	3.7	1.1	0.7	61	23.0
CM	0.9	0.8	0.1	0	4.5*
PM	0.9	0.9	0.1	0	3.4*
Normal Range	3.8–6.8	0.6–1.7	0.4–1.8	60–140	11.5–35.0

* on vitamin E therapy

Figure 2 shows the haplotypes of the chromosomes (deduced from the RFLP analyses) of the parents and children in the J family. Both affected siblings have inherited the apo B gene defined by haplotype B from the father, whereas one child (MJ) has inherited haplotype C and his sibling (SJ) haplotype D from their mother. Similarly in the M family it is possible to distinguish all four parental haplotypes and in this family each child has inherited a different allele from the mother and from the father (not shown). Thus in both families the children have inherited different alleles of the apo B gene from one or both parents. This observation is incompatible with the hypothesis that in these families the mutation causing classical ABL is in or close to the gene for apo B.

Previous immunological studies in patients with ABL have provided conflicting evidence as to whether this condition is caused by a mutation of the apo B gene or from an abnormality of a gene or genes necessary for apo B synthesis and secretion. Thus two groups have failed to detect apo B in intestine[4],[5] which suggests a defect in synthesis, whereas others have reported the presence of apo B in both intestine[6] and liver[6],[7] and of normal sized mRNA in liver[7] which is indicative of a failure of secretion. It is therefore probable that ABL is a heterogeneous disorder that may be caused by defects of genes that are separately involved in regulating synthesis and either the assembly or the secretion of apo B containing lipoproteins

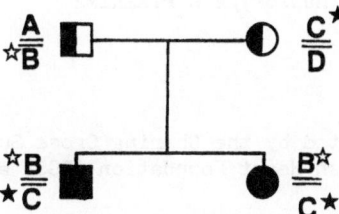

Figure 1. A model pedigree to illustrate co-segregation of a gene with a recessive disorder. In this family parental haplotypes have been unambiguously deduced.

Many of the steps in the synthesis of apo B containing lipoproteins from the liver and the intestine have yet to be elucidated. Pulse-chase experiments, in Hep G2 cells, show that it takes roughly 30 mins for a newly synthesised apo B molecule to be secreted as a lipoprotein[8]. During its synthesis, nascent apo B becomes associated with the endoplasmic reticulum and then passes through the Golgi apparatus. During this time apo B is N-glycosylated and other post-translational events, such as covalent acylation and phosphorylation, take place. There are thus several potential points in this process where mutations may block or alter the rate or production of apo B containing lipoproteins.

An efficient strategy, to map the defect causing recessive disorders by studying the affected offspring of consanguinous marriages, has recently been suggested[9]. This approach may be applicable to determine the gene defect in ABL. It remains to be seen whether in different patients with ABL the defect is in different genes resulting in the same phenotype. The gene or genes, whose functions appear to be vital for the normal secretion of apo B containing lipoproteins, may also be involved in an important step in the normal control of lipoprotein secretion.

The results in our present study provide genetic evidence that in at least some patients ABL may not result from a defect in the gene for apo B. We found no evidence of gross deletions or insertions in the apo B gene from the two families studied nor in six other unrelated ABL patients. Results from the linkage analysis in the families each with two affected children, allows us to rule out the involvement of mutations in or near the apo B gene in both these families.

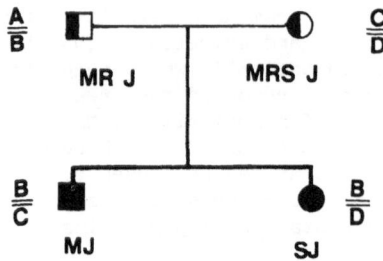

J FAMILY

Figure 2. Pedigree of the J Family. DNA was digested with the enzymes EcoRI (R) XbaI (X) MspI (M) PvuII (P). The unambiguously deduced haplotypes are as follows: Haplotype A P2X1R1M2, haplotype B P1X2R1M2, haplotype C P2X1R2M1 and haplotype D P1X2R1M2.

ACKNOWLEDGEMENTS

This work was supported by the Charing Cross Sunley Research Centre and grants from the British Heart Foundation (RG5) and the MRC.

REFERENCES

1. Kayden HJ. Abetalipoproteinemia. Ann.Rev.Med. 23: 285–296 (1972).
2. Herbert PN, Assmann G, Gotto AM Jr. and Frederickson DS. Familial lipoprotein deficiency; abetalipoproteinemia, hypoabetalipoproteinemia and Tangiers Disease. In 'The Metabolic Basis of Inherited Disease'. JB Stanbury, JB Wyngaarten, DS Fredrickson, JL Goldstein and MS Brown. (Eds) McGraw Hill Inc., New York pp 589–621 (1983).
3. Muller DPR, Lloyd JK and Wolff OH. Vitamin E and neurological function. Lancet i: 225–228 (1983)
4. Glickman RM, Green PH, Lees RS, Lux RE and Kilgore A. Immunofluorescent studies of apolipoprotein B in intestinal mucosa, absence in abetalipo-proteinemia. Gastroenterology 76: 288–292 (1979).
5. Levy E, Marcel YL, Milne RW, Grey VL and Roy CC. Absence of intestinal synthesis of apolipoprotein B-48 in two cases of abetalipoproteinemia. Gastroenterology 93: 1119–1126 (1987).
6. Dullaart RFP, Speelberg B, Schuurman H-J, Milne RW, Havekes LW et al. Epitopes of apolipoprotein B-100 and B-48 in both liver and intestine Expression and evidence of local synthesis in recessive abetalipo-proteinemia. J.Clin.Invest. 78: 1397–1404 (1986).
7. Lackner KJ, Monge JC, Gregg RE, Hoeg JM, Triche TJ, Law SW and Brewer HB Jr. Analysis of the apolipoprotein B gene and messenger ribonucleic acid in abetalipoproteinemia. J.Clin.Invest. 78: 1707–1712 (1986).
8. Olofson S-V, Bjursell G, Bostrom K, Carlsson P, Elovson J, Protter AA, Reuben MA and Bonders G. Apolipoprotein B: structure, biosynthesis and the role in the lipoprotein assembly process. Atherosclerosis 68: 1–17 (1987)
9. Lander ES and Botstein D. Homozygosity mapping: A way to map human recessive traits with the DNA of inbred children. Science 236: 1567–1570 (1987).

RFLPS OF APOB GENE

F. Turturro, J. Heibig, G.C. Ghiselli, and A.M. Gotto, Jr.

The Methodist Hospital, Houston, TX 77030 and
The Vetrans Administration Medical Center, Houston, TX 77032

INTRODUCTION

Epidemiological studies have shown[1-3] that the plasma concentration of LDL cholesterol and apolipoprotein B-100 are positively correlated with the incidence of coronary artery disease. ApoB-100 is believed to have an important role in the alteration of lipid metabolism that may lead to premature atherosclerosis[2,4]. Apolipoprotein B-100 is the molecular determinant of some apoB-containing lipoproteins, serving as a recognition signal for LDL receptor-mediated endocytosis in a variety of cells[5]. More information on the nature of the protein is being obtained through the techniques of molecular biology. In recent years (1985-1986)[6-10] the amino acid sequence of apoB-100 was deduced from the nucleotidic sequences of a number of cDNAs and genomic DNAs, which were isolated from human liver, intestine, and hepatoma libraries. Speculation about the structure-function relationships of the protein is now possible, especially regarding functional domains of apoB-100[8-10]. The apolipoprotein B-100 may contain alterations of structure, and since 1985 several investigators have targeted the gene of apoB for RFLP studies[11-21]. One of the most interesting regions of the protein seems to be the COOH-one third of the protein, because the putative region for receptor binding to the LDL or B/E receptor has been located in this part between residues 3345 and 3381[22].

RFLPS AND APOB GENE

Restriction fragment length polymorphism (RFLP) identifies an altered pattern of the fragments of a given gene, when genomic DNA from different individuals is digested by endonucleases type II. The molecular bases of alterations in the gene are grouped into point mutations and mutations producing sequence rearrangements. The point mutations can occur: 1) in the coding part of the gene; 2) in the intervening sequence; 3) in regulatory sequences; 4) at critical codons; 5) at flanking sequences. Mutations producing sequence rearrangement are the products of deletion, inversion, duplication, and translocation of nucleotidic sequence within the gene or in flanking regions of it. The functional consequences of these alterations on the gene product are so different, ranging from a real mutant (position of the alteration in critical parts of the gene) to silent mutations (wobble position of the codon, or non-functional flanking regions of the gene). The practical use of RFLP is defined by any possible correlation between alterations (functional and silent) of the genetic background and metabolic and/or clinical abnormalities of the gene products (genetic marker). Blackhart et al[23] published data on the structure of the apoB-100 gene. Two major striking features of the apoB gene are: 1) the unusual length of 2 exons for the size limits of a mammalian gene (exon 29, 1906 bp and exon 26, 7572 bp); and 2) the asymmetric distributuion of introns. The gene of apoB-100, which has been localized on the chromosome 2 (region 2 cm → 2 pter) by different investigators[24-28] has a length of 43 Kb with 28 introns and 29 exons. The position of 24 of the 28 introns is in the 5' terminal region of the gene and two long open reading frames (ORFs) are in the 3' portion of the gene (exon 26, 7572 bp and exon 29, 1906 bp). TATA box and CAAT box are respectively located 29 nucleotides 5' of the TATA box.

Genetic and epidemiological studies of the clinical association between RFLPs of the gene of the apoB-100 and alteration of lipid metabolism have been conducted by different groups of investigators in the last three years[29-35]. The endonucleases of type II, which identify RFLP in the 3' region of the apoB-100 gene and in flanking regions and which have been used in those studies, are as follows: EcoRI, Xba I, Msp I, and Hind III. EcoRI RFLP identifies a restriction site polymorphism due to a single base change in the coding region of the gene. A replacement at position 4154 of glutamic acid (common 10.5 Kb allele) by lysine (rare 12.5 Kb allele) produces this polymorphism, the consequence of which on the protein is unknown[30]. Xba I enzyme identifies an RFLP which is due to a single base change in the central part of the coding region, and results in no alteration of the amino acidic sequence (threonine 2488) in the protein, because of the third position of the mutation in the codon[30,31]. An insertion-deletion mechanism of repetitive AT sequence

Table 1 -- ApoB-100 RFLPs Associated with Clinical Abnormalities

Reference/ Authors	Enzyme/ Allele	RFLP Type	Clinical and/or Metabolic Abnormalities	Probe
(29) A. Law et al 1986	Xba I 5.0 Kb (X1) 8.6 Kb (X2)	RSP	X1X1 and X1X2 association with higher TG and CH levels vs. X2X2	(ABI + 7)
(30) R.A. Hegele et al 1986	Xba I 8.6 Kb (X1) 5.0 Kb (X2) 4.4 Kb (X3)	RSP	X1, R1 and ID1 association with M.I. cases vs. controls (P < 0.01)	pB 23
	EcoRI 13.1 Kb (R1) 11.0 Kb (R2) 11.0 Kb (R2) 2.1 Kb (R2)	RSP	No association with variation of LDLC or apoB	pB8–pB27 both pB8 pB27
	Msp I $\geq$ 2.5 Kb (ID1) < 2.5 Kb (ID2)	ID		pB8
(31) K. Berg et al 1986	Xba I 5.0 Kb (X1) 8.6 Kb (X2)	RSP	Linkage disequilibrium between X2 and the Ag(x) antigenic determinant	5.07 Kb (T_2 thrombin peptide coding sequence + 270 bp of 3' untranslated region)
	EcoRI 14.1 Kb (R2) 12.0 and 2.1 Kb (R1)	RSP	Weak association with the Ag(x) antigen	As above
	Msp I 2.2–2.8 Kb	ID	No association	As above
(32) Y. Ma et al 1987	Xba I 5.0 Kb (X1) 8.0 Kb (X2)	RSP	Imperfect association with the Ag (g/c) (14/17)	AB6–AB14 (4.6 Kb – 3' region)
	EcoRI 12.0 Kb and 2.0 Kb (R1) 14.0 Kb (R2)	RSP	Strong association with the Ag (t/z) (17/17)	As above
(33) P.J. Talmud et al 1987	Xba I 8.5 Kb (X1) 3.5 Kb (X2)	RSP	X2 association with type III vs. normolipidemic cases. X2 association with CH and TG levels in normolipidemic cases	3.5 Kb (EcoRI unique fragment)
	EcoRI 10.5 Kb (R1) 12.45 Kb (R2)	RSP	No correlation	pAB3
	Msp I 2.6 Kb (M1) 2.3 Kb (M2) 2.4 Kb (M2)	ID	No correlation	pAB3
(34) K. Jenner et al 1988	Xba I (X$^+$, X$^-$) EcoRI (E$^+$, E$^-$)	RSP RSP	Linkage disequilibrium of X$^+$ with a genetic locus for high TG levels vs. X$^-$; no statistically significant difference in normo- and hyperlipidemic cases	pB2, pB3, pB4 (3.75 Kb)
	E$^+$ X$^+$ (0.68) E$^+$ X$^-$ (0.30) E$^-$ X$^+$ (0.02) E$^-$ X$^-$ (0.00)		Random association of haplotypes in a normolipidemic population	
	HVE (5 alleles) Hind III/Sst I 1.3 ± 0.2 Kb	ID	Informative locus for hyperlipidemic families?	
(35) M.V. Monsalve et al 1988	Xba I 8.5 Kb (X1) 3.5 Kb (X2)	RSP	X1 R2 haplotype association with vascular diseases (coronaries, carotid and peripheral arteries) vs. controls	3.5 Kb (EcoRI unique fragment)
	EcoRI 10.5 Kb (R1) 12.5 Kb (R2)	RSP	No significant correlation with any of the 3 different parts of the vascular system in the cases	pAB3
			X1 X1 association with low CH level vs. X2 X2 high CH level. (Peripheral arterial disease); reversed association in coronary-carotid disease	

downstream of the 3' region of the gene (untranslated region), is the molecular basis of Msp I and Hind III RFLPs, and produces no apparent alteration of the protein[17,19,32]. Table 1 summarizes the major studies of RFLPs of the apoB-100 gene and clinical abnormalities.

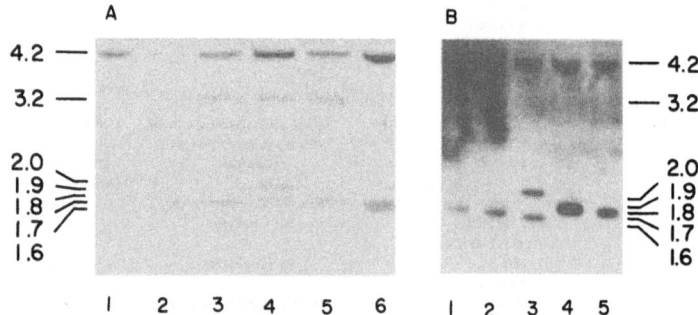

Figure 1. Hind III RFLP, Southern blotting from different subjects.

HIND III RFLP OF APOB GENE

In this study we examined apoB-100 polymorphism for potential corre- lations with abnormalities in the metabolism of the apolipoprotein in subjects undergoing coronary angiography at the Veterans Administration Medical Center in Houston, Texas.

DNA was extracted from the buffy coats of frozen whole blood samples of 53 male caucasians in their 50's and 60's. The DNAs of these subjects were digested by several type II endonucleases. Separation and resolu- tion of the fragments, resulting from complete digestion of genomic DNA, were carried out by electrophoresis on a standard horizontal agarose gel (.8% - 1% Agarose). Southern blotting on nylon membranes was performed, and after that the filters were hybridized with a P^{32} labeled probe. The probe (B1-8) is a 1.5 Kb fragment of the 3' region of cDNA subcloned in pBR322, is derived from a human liver library cloned in λ gt 11, and was a kind gift from Dr. L. Chan. The DNAs were digested by EcoRI, Xba I, Hind II, and Bam HI and no polymorphism was detected. Hind III identi- fied a complex pattern of fragments with length in a range from 1.6 Kb to 2.0 Kb. The frequency of 5 possible alleles was: 1.6 Kb (0.13), 1.7 Kb (0.43), 1.8 Kb (0.22), 1.9 Kb (0.02), and 2.0 Kb (0.20) (Figure 1).

After the initial results among these first 53 subjects and the definition of the RFLP, we started to screen a larger population of CAD subjects, normo- and dyslipidemic. For each subject we are taking a clinical history (regarding major risk factors for CAD and familiarity), measuring weight and height, collecting blood samples for the determination of cholesterol, triglycerides, HDLC, and apolipoprotein B levels. Preliminary data have shown that subjects with the 1.8 Kb Hind III allele present a tendency to higher values of cholesterol compared to the other alleles (1.6, 1.7, 1.9, 2.0 Kb) (Data not shown).

SUMMARY

Xba I and EcoRI RFLPs (RSP) have been reported to be statistically significant in being positively or negatively correlated with cholesterol and triglyceride levels in normolipidemic and hyperlipidemic populations, in their association with coronary, carotid and peripheral arterial disease; and in their use in linkage studies. Msp I RFLP (ID) has a low content of statistical information. HVE Hind III-Sst 1 RFLP (ID) may have an informative locus for the study of hyperlipidemic families[32]. The allele 1.8 Kb of Hind III RFLP in our study seems to correlate with the tendency to higher levels of cholesterol. These are very preliminary data and they need to be confirmed by a larger population and by the definition of the molecular nature of the RFLP. Both of these studies are underway.

REFERENCES

1. I. Stamler, Population Studies, in: "Nutrition, Lipids and Coronary Heart Disease," R. Levy, B. Rifkind, B. Dennis, and N. Ernse, ed., Raven Press, New York (1979).
2. T. F. Whayne, P. Alaupovic, M. D. Curry, E. T. Lee, P. S. Anderson, and E. Schechter, Plasma apolipoprotein B and VLDL, LDL, HDL-Cholesterol as risk factors in the development of coronary artery disease in male patients examined by angiography, Atherosclerosis 39:411 (1981).
3. Lowering blood cholesterol to prevent heart disease, Consensus Conference, JAMA 253:2080 (1985).
4. P. Avogaro, G. B. Bon, G. Cazzolato, G. B. Quinci, and F. Belussi, Plasma levels of apolipoprotein A-I and apolipoprotein B in human atherosclerosis, Artery 4:385 (1978).

5. M.S. Brown and J.L. Goldstein, A receptor-mediated pathway for cholesterol homeostasis, Science 232:34 (1986).

6. S. H. Chen, C. Y. Yang, P. F. Chen, D. Setzer, M. Tanimura, W. H. Li, A.M. Gotto, and L. Chan, The complete cDNA and amino acid sequence of human apolipoprotein B-100, J. Biol. Chem. 261:12918 (1986).

7. T. J. Knott, R. J. Pease, L. M. Powell, S. C. Wallis, S. C. Roll, T. L. Innerarity, B. Blackhart, W. H. Taylor, Y. Marcel, R. Milne, D. Johnson, M. Fuller, A. J. Lusis, B. J. McCarthy, R. W. Mahley, B. Levy-Wilson, and J. Scott, Complete protein sequence and identification of structural domains of human apolipoprotein B, Nature 323:734 (1986).

8. C.Y. Yang, S.H. Chen, S.H. Gianturco, W.A. Bradley, J.T. Sparrow, M. Tanimura, W. H. Li, D. A. Sparrow, H. DeLoof, M. Rosseneu, F. S. Lee, Z. W. Gu, A. M. Gotto, and L. Chan, Sequence, structure, receptor binding domains and internal repeats of human apolipoprotein B-100, Nature 323:738 (1986).

9. S. W. Law, S. M. Grant, K. Higuchi, H. Hospattankar, K. Lackner, N. Lee, and H. B. Brewer, Jr., Human liver apolipoprotein B-100 cDNA: Complete nucleic acid and derived amino acid sequence, Proc. Natl. Acad. Sci. USA 83:8142 (1986).

10. C. Cladaras, M. Hadzopoulos-Cladaras, R. T. Notte, D. Atkinson, and V. I. Zannis, The complete and structural analysis of human apolipoprotein B-100: Relationship between apoB-100 and apoB-48, EMBO J 5:3495 (1986).

11. L. Priestley, T. Knott, S. Wallis, L. Powell, R. Pease, A. Simon, and J. Scott, RFLP for the human apolipoprotein B gene: I; Bam HI, Nucl. Acids. Res. 13:6789 (1985).

12. L. Priestley, T. Knott, S. Wallis, L. Powell, R. Pease, and J. Scott, RFLP for the human apolipoprotein B gene: II; EcoRI, Nucl. Acids Res. 13:6790 (1985).

13. L. Priestley, T. Knott, S. Wallis, L. Powell, R. Pease, and J. Scott, RFLP for the human apolipoprotein B gene: III, EcoRV, Nucl. Acids Res. 13:6791 (1985).

14. L. Priestley, T. Knott, S. Wallis, L. Powell, R. Pease, and J. Scott, RFLP for the human apolipoprotein B gene: IV, Msp I, Nucl. Acids Res. 13:6972 (1985).

15. L. Priestley, T. Knott, S. Wallis, L. Powell, R. Pease, H. Brunt, and J. Scott, RFLP for human apolipoprotein B gene: V, Xba I, Nucl. Acids Res. 13:6973 (1985).

16. P. M. Frossard, P. A. Gonzalez, A. A. Protter, R. T. Coleman, H. Funke, and G. Assman, Pvu II RFLP in the 5' of the human apolipoprotein B gene, Nucl. Acids Res. 14:4373 (1986).

17. C. Darnfors, J. Nilsson, A. A. Protter, P. Carlsson, P. J. Talmud, S. E. Humphries, J. Whalstrom, O. Wiklund, and G. Bjursell, RFLPs for the human apolipoprotein B gene: Hind II and Pvu II, Nucl. Acids Res. 14:7135 (1986).

18. T. J. Knott, S. C. Wallis, R. J. Pease, L. M. Powell, and J. Scott, A hypervariable region 3' to the human apolipoprotein B gene, Nucl. Acids Res. 14:9215 (1986).

19. L. S. Huang and J. L. Breslow, A unique AT-rich hypervariable minisatellite 3' to the apoB gene defines a high information restriction fragment length polymorphism, J. Biol. Chem. 262:8952 (1987).

20. F. Turturro, J. Heibig, L. Chan, A. M. Gotto, and G. C. Ghiselli, A complex Hind III polymorphism in the human apolipoprotein B gene (ApoB), Nucl. Acids Res. 15:9618 (1987).

21. L. S. Huang, J. deGraaf, and J. L. Breslow, ApoB gene Msp I RFLP in exon 26 changes amino acid 36 II from Arg to Gln, J. Lipid Res. 29:63 (1988).

22. T. L. Innerarity, K. H. Weisgraber, S. C. Rall, Jr., and R. W. Mahley, Functional domains of apolipoprotein E and apolipoprotein B, Acta Med. Scand. Suppl. 715:51 (1987).

23. B. D. Blackhart, E. M. Ludwig, V. R. Pierotti, L. Caiati, M. A. Onasch, S. C. Wallis, L. Powell, R. Pease, T. J. Knott, M. L. Chu, R. W. Mahley, J. Scott, B. J. McCarthy, and B. L. Wilson, Structure of the human apolipoprotein B gene, J. Biol. Chem. 261:15364 (1986).

24. L. Chan, P. Van Tuinen, D. H. Ledbetter, S. P. Daiger, A. M. Gotto, and S. H. Chen, The human apolipoprotein B-100 gene: A highly polymorphic gene that maps to the short arm of chromosome 2, Biochem. Biophys. Res. Commun. 133:248 (1985).

25. S. W. Law, K. J. Lackner, A. V. Hospattankar, J. M. Anchors, A. Y. Sakaguchi, S. L. Naylor, and H. B. Brewer, Jr., Human apolipoprotein B-100: Cloning, analysis of liver mRNA and assignment of the gene to chromosome 2, Proc. Natl. Acad. Sci. USA 82:8340 (1985).

26. S. S. Deeb, C. Disteche, A. G. Motulsky, R. V. Lebo, and Y. W. Kan, Chromosomal localization of the human apolipoprotein B gene and detection of homologous RNA in monkey intestine, Proc. Natl. Acad. Sci. USA 83:419 (1986).

27. L. S. Huang, D. A. Miller, G. A. P. Bruns, and J. L. Breslow, Mapping of the human apoB gene to chromosome 2p and demonstration of a two-allele restriction fragment length polymorphism, Proc., Natl. Acad. Sci. USA 83:644 (1986).

28. N. Barni, P. J. Talmud, P. Carlsson, M. Azoulay, C. Darnfors, D. Harding, O. Weil, K. H. Grzeschik, G. Bjursell, C. Junien, R. Williamson, and S. E. Humphries, The isolation of genomic recombinants for the human apolipoprotein B gene and the mapping of three common DNA polymorphisms of the gene - a useful marker for human chromosome 2, Hum. Gen. 73:313 (1986).

29. A. Law, L. M. Powell, H. Brunt, T. J. Knott, D. G. Altman, J. Rajput,S. C. Wallis, R. J. Pease, L. M. Priestley, J. Scott, G. J. Miller, and N. E. Miller, Common DNA polymorphism within coding sequence of apolipoprotein B gene associated with altered lipid levels, Lancet 1:1301 (1986).

30. R. A. Hegele, L. S. Huang, P. N. Herbert, C. B. Blum, J. E. Buring, C. H. Hennekens, and J. L. Breslow, Apolipoprotein B gene DNA polymorphisms associated with myocardial infarction, N. Engl. J. Med. 315:1509 (1986).

31. K. Berg, L. M. Powell, S. C. Wallis, R. Pease, T. J. Knott, and J. Scott, Genetic linkage between the antigenic group (Ag) variation and the apolipoprotein B gene: Assignment of the Ag locus, Proc. Natl. Acad. Sci. USA 83:7367 (1986).

32. Y. Ma, V. N. Schumaker, R. Butler, and R. S. Sparkes, Two DNA restriction fragment length polymorphisms associated with Ag (t/z) and Ag (g/c) antigenic sites of human apolipoprotein B, Arteriosclerosis 7:301 (1987).

33. P. J. Talmud, N. Barmi, A. M. Kessling, P. Carlsson, C. Darnfors, G. Bjursell, D. Galton, V. Wynn, H. Kirk, M. R. Hayden, and S. E. Humphries, Apolipoprotein B gene variants are involved in the determination of serum cholesterol levels: A study in normo- and hyperlipidemic individuals, Atherosclerosis 67:81 (1987).

34. K. Jenner, A. Sidoli, M. Ball, J. R. Rodriguez, F. Pagani, G. Giudici, C. Vergani, J. Mann, F. E. Baralle, and C. C. Shoulders, Characterization of genetic markers in the 3' end of the apoB gene and their use in family and population studies, Atherosclerosis 69:39 (1988).

35. M. V. Monsalve, R. Young, J. Jobsis, S. A. Wiseman, S. Dhamu, J. T.
     Powell, R. M. Greenhalf, and S. E. Humphries, DNA polymorphisms of
     the gene for apolipoprotein B in patients with peripheral arterial
     disease, <u>Atherosclerosis</u> 70:123 (1988).

APOLIPOPROTEIN B GENETIC DIFICIENCES

R. Infante

I.N.S.E.R.M. Hospital Saint Antoine 75012 Paris - France

Abeta and hypobetalipoproteinemias designate an heterogeneous group of syndromes characterized by a marked decrease or absence of apoprotein B (apo-B)-containing lipoproteins in fasting and postprandial serum. Because of the essential role of apo-B containing lipoproteins in triglyceride and cholesterol transport from the liver and intestine into the circulation, defective synthesis and/or secretion of apo-B is commonly associated with liver steatosis and/or intestinal fat malabsorption and fat accumulation in the enterocytes. In addition, intestinal and plasma transport of fat soluble vitamins (A, E, D, K) may be impaired which can lead to retinal degeneration and demyelinizing lesions of the central and peripheral nervous system.

The original (and long accepted) concept of abetalipoproteinemia as a phenotypic expression of lack of apo-B synthesis has rapidly changed in recent years due to the increasing utilisation of the techniques of modern cell and molecular biology. Also, the cloning of human apo-B, the discovery of two forms (apo-B100, apo-B48) of post-transcriptional expression of a single gene, as well as the sequencing of apo-B and the utilisation of monoclonal antibodies have stimulate investigators to better define the pathogenesis and molecular defects involved in these genetic diseases. Other apolipoprotein deficiencies or mutations benefit from the same approach.

Clinical manifestations (Table I) - The earliest and most characteristic manifestations of abeta and some forms of hypobetalipoproteinemia are diarrhea and steatorrhea due to intestinal fat malabsorption (1). Very often children are referred to hospital with chronic diarrhea and failure to thrive. When maternal milk or free diets are replaced by low fat diets, this symptom regresses rapidly, and the patients recover normal growth following diets that contain less than 5 g fat/day. Although the clinical picture particularly in early infancy, resembles that of coeliac disease, diagnosis can be established by the whitish colour of the intestinal mucosa and by the histologically normal shape and height of the intestinal villi and epithelial cells. Biopsies of the latter show them filled with fat droplets after a fatty meal and sometimes also after fasting biopsies.
If diagnosis is not established in infants and an appropriate diet is not prescribed, fat and vitamin malabsorption with result in physical

TABLE I

CLINICAL SYMPTOMS IN APO B DEFICIENCIES

	GROWTH RETARDATION	MENTAL RETARDATION	NEUROLOGICAL IMPAIRMENT	RETINITIS PIGMENTOSA	ACANTHOCYTOSIS
ABETALIPOPROTEINEMIAS	YES	YES	YES	YES (>8 yr)	YES
FAM. HYPOBETA	-	NO⟶YES	YES (> 8 yr)	NO⟶YES	YES
NORMO TG ABETA	NO/YES	YES	YES/NO	NO	NO/YES
APO B100 DEFICIENCY	NO	NO	NO (> 2 yr)	NO	FEW
ANDERSON'S DISEASE	YES	YES/NO	YES/NO	NO	NO
HYPOBETA B37	NO	NO	NO	NO	NO
HYPOBETA-VLDL-HYPERCATABOLISM	NO	NO	NO	YES	NO

96

and mental retardation and in progressive impairment of neurological and retinal functions. Thus, early detection of the disease is important and it is necessary to initiate dietary management in order to prevent neuromuscular and retinal disfunction. This can be achieved on the basis of family history, physical examination, intestinal biopsies, the presence of chronic diarrhea or loose stools with an abnormal fat content, and a decrease or absence of serum apo-B containing lipoproteins, low cholesterol and triglyceride levels and sometimes moderate hypochromic anemia and acanthocythosis (Table II).

However, this early clinical picture applies to patients with abetalipoproteinemia, the homozygous form of familial hypobetalipoproteinemia and Anderson's disease (1-5). Other apobetalipoprotein genetic disorders may result in isolated retinitis, malabsorption, or retardation, or they may be clinically silent and because of normal intestinal absorption do not interfere with development. In such cases, plasma lipid and apolipoprotein analysis may provide valuable data for diagnosis (6-12). Gastrointestinal manifestations turn particular attention to intestinal endoscopic and histological examinations. Liver needle biopsy is useful, since the impairment of lipoprotein synthesis or processing leads to moderate or severe fat accumulation in the hepatocytes and to hepatomegaly. Liver steatosis has been found in most of the apo-B-deficient patients when liver scanning, liver endoscopy or needle biopsies have been performed. It is unclear whether liver steatosis reflects other metabolic disturbances related to essential fatty acids or fat-soluble vitamins deficiencies, besides the impairment of lipid transport.

Nutritional consequences (Table III) – Most of the symptoms observed in abeta or hypobetalipoproteinemic patients results from intestinal malabsorption. Fat-rich maternal milk or free diets are poorly tolerated. Vomiting and chronic diarrhea, primarily due to fat malabsorption, starve the patients, not only of fatty acids but also of carbohydrates and proteins. Since digestion absorption and transport of these latter nutrients are not primarily affected in these diseases feeding a low fat diet (usually less than 5 g/day for children and 10 g/day for adults)) improves diarrhea and allows normal absorption of proteins, carbohydrates and enough essential fatty acids for normal development. Low fat diets, however, do not resolve fat soluble vitamin deficiency. It has been shown that in such patients, intraluminal fat digestion, micelle formation and fatty acid absorption by enterocytes are normal, provided that the amount of dietary fat does not induce diarrhea. Fatty acids are also normally esterified by the enterocytes to triglycerides and phospholipids. Thus, in these conditions, vitamins are normally absorbed but, like triglycerides, they are poorly transported to the lymph and portal circulation (Fig 1).

Vitamin A is esterified in the enterocyte to retinol palmitate which is incorporated in the endoplasmic reticulum and Golgi apparatus to chylomicrons and secreted into the intercellular space lymphatic circulation. Normally, some retinol palmitate can be transported directly to the circulation by a specific carrier protein (retinolbinding protein RBP). Presumably, in abeta and hypobetalipoproteinemic patients with a block in chylomicron production or secretion, small amounts of vitamin A are absorbed via RBP. However, the transport capacity of RBP seems to be limited, since partial restoration of normal plasma or tissue levels can be only obtained after administration of high oral (or parenteral) doses of vitamin A.

Vitamin K and E can be also transported by carrier proteins, though under normal conditions chylomicrons are probably the main form of transport. Vitamin D deficiencies have not been reported in abeta

VITAMINS TRANSPORT

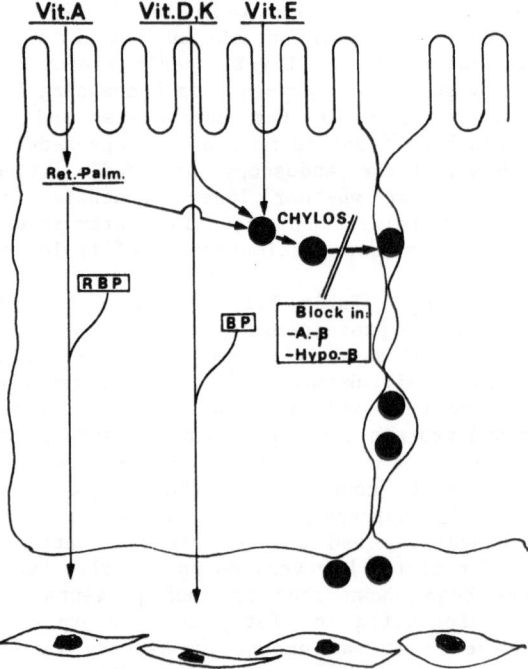

Fig 1. Transport of fat soluble vitamins in the absorptive intestinal cell.
Intracellular binding proteins (B, P) carry out a fraction of vitamins A, D and K, but, in normal condition the most part of vit.
A, D and K and all vit. E is likely transporting into the lymphatic
circulation by chylomicrons.

and hypobetalipoproteinemias possibly since a poor absorption is compensated by endogenous tocopherol synthesis in the skin. However, haemorrhagic accidents due to vitamin K deficiency have been reported.

Among the metabolic consequences of fat malabsorption, tocopherol deficiency is certainly responsible for the severe neuromuscular and retinal lesions characteristic of the advanced stages of abetalipoproteinemia. Many arguments are in favour of this assumption : a) In experimental animals vitamin E deficient diets produce demyelinisation, axonal degeneration, myopathy and retinal degeneration and pigmentation; b) plasma and tissue tocopherol levels are constantly very low and can be normalized in tissues only after administration of grams of vitamin E per day; c) early tocopherol supplementation in infants prevents most of the neuroretinal symptoms, and d) massive vitamin E administration stops or partly reverses the already present neuromuscular degeneration. Since vitamin E is mainly transported to tissues bound to apo-B containing lipoproteins, its plasma level rarely exceedes 10-20 % of normal values after long-term tocopherol therapy. Thus, tocopherol determination in adipose tissue biopsies is a more reliable index to assess its concentration in other tissues (nervous system, muscle, liver, etc) (2).
Besides fat soluble vitamins malabsorption of other essential nutrients (essential fatty acids, aminoacids) in abeta and hypobetalipoproteinemic patients contributes to the development of the neuromuscular syndrome, as well as to physical and mental retardation. Thus, it is important to control diarrhea which normalizes carbohydrate and aminoacid malabsorption, by maintaining a fat (preferentially unsaturated vegetable oils) intake of 5-10 g/day or more. Depending on individual tolerance, enough essential fatty acid can be absorbed and transported into the circulation, in spite of defective chylomicron secretion, perhaps by exocytosis of phospholipid or phospholipid-apo-AI (AIV ?) complexes. Another interesting source of polyunsaturated fatty acids in these patients is the periodical intraveinous infusion of lipid emulsions. Medium chain triglycerides can be used as absorbable fat to correct malnutrition. However, longterm administration must be avoided because of the risk of liver fibrosis (13).

Cholesterol homeostasis in apo-B deficiencies – Cholesterol synthesis in many mammalian cells is regulated by the amount of cholesterol taken up from the plasma or extracellular fluid, in vivo, or from the medium in cultured cells, in vitro. Lipoprotein-cholesterol uptake is mainly, but not only, mediated by endocytosis of particles containing apo-B or apo-E, via a high affinity membrane receptor (apo-B, E or LDL receptor) (14). Depletion of the culture medium in vitro of apo-B-containing lipoproteins results in maximal expression of this receptor and in derepression of the key enzyme in cholesterogenesis HMG CoA reductase. This can be reversed by the addition of LDL. Abeta or hypobetalipoproteinemia may mimic this situation because of low (or absent) apo-B and low plasma cholesterol levels and thus, both an increased number of LDL receptors and a high rate of cholesterol synthesis may be predicted. Though in vitro studies of cholesterol synthesis in intestinal and liver cells and LDL-receptor activity in liver cell membranes from abetalipoproteinemia patients have not been published, some indirect evidence has been obtained from determination of cholesterol synthesis rate in skin biopsies and freshly obtained blood mononuclear cells, incubated with radioactive acetate or tritiated water. Former studies (15,16) indicated an increased cholesterogenesis in abetalipoproteinemia. However, other in vitro (17) and vivo studies (18-20) indicate that cholesterol synthesis and LDL-receptor expression are repressed in spite of the absence of apo-B and LDL-cholesterol. Presumably, in apo-B deficiency, HDL_2 which

TABLE III

MORPHOLOGICAL CHANGES IN APO B DEFICIENCIES

| | INTESTINE | | LIVER | |
| | ENDOSCOPY | HISTOLOGY | | HISTOLOGY |
	NORMALLY SHAPED WHITISH VILLI	NORMAL ENTEROCYTES LIPID VACUOLES	HEPATOMEGALY	STEATOSIS
ABETALIPOPROTEINEMIAS	YES	YES	YES or NO	YES
FAM. HYPOBETA	YES	YES	-	YES
NORMO TG ABETA	NO/ ?	NO/ ?	? / ±	?
APO B100 DEFICIENCY	NO	YES	YES	YES
ANDERSON'S DISEASE	YES	YES	YES/NO	YES
HYPOBETA B37	NO/ ?	?	NO	?
HYPOBETA-VLDL-HYPERCATABOLISM	NO	NO	?	?

contain as much as twice apo-E and cholesterol as HDL_2 from normal
plasma compensate the absence of LDL for both cholesterol transport
between the liver and peripheral tissues and for down regulation of
the LDL (apo-B, E) receptors (21,22). Sterol balance studies indicate
an increased fecal sterol output in some patients (18,19,20,23).
Cholesterol excretion in patients with fat malabsorption may express
a decrease in bile cholesterol reabsorption, which could explain increased
fecal cholesterol losses. This (and eventually bile acid malabsorption
due to diarrhea) should stimulate endogenous cholesterol synthesis
in the liver to reach a new steady state of cholesterol homeostasis.
In these conditions, the metabolic status (cholesterogenesis, apo-B
receptors) of blood or peripheral cells does not necessarily reflect
the situation of the two main organs (liver, intestine) involved in
cholesterol synthesis.
Incidentally, the relative "normalisation" of cholesterol homeostasis
by HDL_2 in abetalipoproteinemic patients could not be complete in the
adrenals and the ovary (24,25,26) two tissues that, like the testis,
synthesize steroid hormones mainly from cholesterol delivered by LDL
via the apo-B, E receptor (27).

Apoprotein B in plasma and tissues – Table IV depicts the plasma
concentration and the presence of apo-B in the liver and intestine.
Apo-B is undetectable by chemical and immunological techniques (ELISA)
in recessive abeta and in the homozygous form of familial
hypobetalipoproteinemia; the heterozygous patients of the latter syndrome
have about 50 % of the normal levels of apo-B, whereas in the former
the levels are indistinguishable from normal subjects. This feature
facilitates the differential diagnosis between abeta and hypobeta
homozygous patients who present similar clinical and biological
profiles.
In abetalipoproteinemia, familial hypobeta and Anderson's disease, HDL
and apo-AI plasma concentrations are usually low. Although there is
no explanation for this decrease, it is likely that a defective
production of intestinal apo-A, due to lack of chylomicron secretion,
may lead to low levels of HDL, which is partly formed in the circulation
by enrichment of nascent HDL with apo-AI and other surface material
derived from chylomicron catabolism. Apo-AIV, another chylomicron bound
intestinal apoprotein, is low in both fasting and postprandial serums.
Apo-CIII$_1$ deficiency has been described in abetalipoproteinemia. Apo-E,
which originates in the liver, is increased in HDL in complete apo-B
deficiencies and also sometimes in partial apo-B deficiencies, which, as
discussed below, may play a role in maintaining cholesterol homeostasis.
In Anderson's disease (9,10) the lack of chylomicron secretion is
responsible for the absence of apo-B48 in fasting and postprandial serum.
However, it is unclear why these patients have about 50 % of the normal
levels of apo-B100 which is mainly, if not exclusively, synthesized
in the liver. Normotriglyceridemic abetalipoproteinemia is almost the
mirror reflection of Anderson's disease. Intestinal fat transport
and chylomicron secretion are normal and apo-B48 can be detected in
postprandial serum (6,7). However, apo-B100 is absent from the
circulation, suggesting a defective synthesis or secretion of apo-B
containing lipoproteins by the liver. Other forms of
hypobetalipoproteinemia (8,11,12) are associated with hypercatabolism
of a presumably abnormal B100 or with a more complex combination of
genetic errors (hypobeta B37). In the latter group some members have
low levels of apparently normal apo-B100, others have an abnormal,
truncated form of apo-B (B37) and others, compound heterozygotes,
have both the abnormal B37 and low levels of B100 (29). Investigations of
apo-B at the cellular level have not been carried out in many of these
patients.
In recessive abetalipoproteinemia, apo-B can be detected in intestinal

TABLE III

NUTRITIONAL AND BIOCHEMICAL CONSEQUENCES OF APO B DEFICIENCY

	INTESTINAL MALABSORPTION LIPIDS			
	STEATORRHEA	VIT. A-D-K	VIT. E	ESSENTIAL FA
ABETALIPOPROTEINEMIAS	YES	YES/NO	YES	YES/NO
FAM. HYPOBETA	YES	YES/NO	YES	YES/NO
NORMO TG ABETA	NO	NO	YES	NO
APO B100 DEFICIENCY	YES	?	YES	?
ANDERSON'S DISEASE	YES	YES/NO	YES	YES/NO
HYPOBETA B37	NO	NO	NO	NO
HYPOBETA-VLDL-HYPERCATABOLISM	NO	LIKELY (A)	LIKELY	NO

or liver biopsies in some patients (30,31) but not in others (32,33),
using immunoperoxidase staining techniques. A negative reaction, however,
cannot be interpreted as a reliable criterium of the lack of apo-B
synthesis; the protein may exist for the concentrations below the
detection limits of the immunochemical reaction and, on the other hand,
the apoprotein may be synthesized and catabolized in the intestinal
or hepatic cells. In a small number of abetalipoproteinemic patients
in which it has been investigated, apo-B mRNA has been always found
in normal or elevated amounts. It may also be predicted that in some
cases of abeta or hypobetalipoproteinemia, an apo-B of abnormal
structure is synthesized, secreted and very rapidly hydrolized in the
circulation or by the cells.

Molecular biology of abeta   and hypobetalipoproteinemias - The
apoprotein B cloning and sequencing has not been establish that in 1986.
The gene mRNA structures have been elucidated and cDNA probes have been
available, thus providing new tools to study the apo-B genetic defects
at the molecular level (34,35).
The human apo-B gene is unusually long (43 Kb), contains 29 exons and
28 introns and has been assigned to chromosome 2. This single gene
directs the synthesis of two apoprotein B forms, B100 and B48, through
a post transcriptional substitution of a single nucleotide in the mRNA
chain which transforms the glutamine codon (CAA) at residue 2153 to
an in frame stop codon (UAA)(36-38).

This results in the complete translation of the mRNA coding for
apoprotein B100 (512 KDa) or the translation of a mRNA containing the
stop codon leading to the production of apo-B48. In fact, apo-B48 (48
% of the aminoterminal chain of B100) might be synthesized by two
mRNAS, one of 14.1 Kb, containing the stop codon and a second of 7.5
Kb completed after the stop codon by a poly A tail. Though both stop
codon mRNAS have been identified in human liver and intestine biopsies
together with the apo-B100 14.1 Kb mRNA, more direct proofs are needed
on the in vivo expression of mRNA-apoB100 in the intestine and of
those of B48 in the liver. B100 synthesis and in vitro secretion by
human intestinal explants has been found by some authors but not by
others.
Studies of DNA polymorphism in abetalipoproteinemia patients have shown
that the apo-B gene does not contain major insertions or deletions which
could explain an abnormal apo-B structure. Moreover the corresponding
mRNA is always normal in size. Although point mutations could produce
modifications in important domains (lipid binding, apo-B receptor
binding) which could block the chylomicron or VLDL assembly or increase
the apo-B catabolic rate these abnormalities have not as yet been
documented.

In abetalipoproteinemia, apo-B mRNA has also been found in intestinal
biopsies in a normal or elevated number of copies and apoprotein B,
revealed by immunoperoxidase staining of enterocytes was normal or
decreased (32,39). In some patients, however, the same technique failed
to show any apo-B in the cells, suggesting that apo-B mRNA is either
not functional or that the apoprotein is synthesized but rapidly degraded
in the cell, perhaps as a result of structural defects in the apoprotein.
Other possible causes, such as defective acylation, phosphorylation
or glycosilation of apo-B have been proposed in these cases but have
never been demonstrated. Technical reasons make this approach very
difficult in abetalipoproteinemia, because of the absence of apo-B
in plasma, thus limiting the study of an abnormal apo-B to the minute
amounts which can be extracted from liver or intestinal biopsies.
Linkage analysis of some familial hypobeta kindred has suggested
that the molecular defect in these patients might be a mutation in the

TABLE IV

APOLIPOPROTEIN B DEFICIENCIES

| | | APOPROTEIN B | | | |
| | | PLASMA | | TISSUES | |
	TRANSMISSION	B100	B48	LIVER	INTESTINE
ABETALIPOPROTEINEMIAS	Auto. recessive	Absent	Absent	Present or Absent	Absent
FAM. HYPOBETA	Auto. dominant	Homoz. Hetero.	Absent Normal or low	Present	Present
NORMO TG ABETA	?	Absent	Normal	?	
APO B100 DEFICIENCY	Auto. recessive	Absent	Low	?	
ANDERSON'S DISEASE	Auto. recessive	Low	Absent	Present	Present
HYPOBETA B37	Auto. dominant	Low	Low	?	
HYPOBETA-VLDL-HYPERCATABOLISM	?	Low	?	?	

apo-B gene (40). This was confirmed by recent reports on homozygous hypobetalipoproteinemia patients showing mutations in apo-B gene exons (39,41-43). In two patients (43) a deletion of a single nucleotide at codon 1794 resulted in the synthesis of a truncated apoprotein (apo-B39) which is present in small amounts in circulating VLDL and LDL; the second patient had a mutation at codon 1306 resulting in a stop codon but the predicted truncated apo-B was not found in plasma. In kindred (42) with low levels of apo-B100 and truncated apo-B containing the N terminal 1728 aminoacids of B100,it has been found that patients bearing the apo-B37 allele have a 4 base pairs deletion at the exon 26,resulting in a frameshift and one novel aminoacid substitution (valine) followed by a stop codon.

The molecular defects in Anderson' disease and other forms of hypobetalipoproteinemias are unknown. The use of recent techniques of molecular biology have revealed the heterogeneity of molecular defects which can be expressed by closely-related phenotypes. It can be predicted that, in the futur, other apo-B related genetic defects should be discovered. These could result from gene defects or transcriptional modifications, abnormal mRNA processing, decreased translation, cotranslational or post-translational defects. Mutant forms of apo-B may have poor lipid binding properties or undergo rapid degradation.

However, the presence of low levels of other circulating apolipoproteins in abeta and hypobetalipoproteinemias do not have, as yet, a rational explanation. One appealing hypothesis is that in some patients apolipoprotein gene expression is normal while other proteins, essential for intracellular lipoprotein traffic and secretion processes are lacking or non functional.

REFERENCES

1. Herbert, P.N., Assmann, G., Gotto, A.M., Fredrickson, D.S. Familial lipoprotein deficiency abetalipoproteinemia, hypobetalipoproteinemia, and Tangier disease. In, the Metabolic Basis of Inherited Diseases. J. B. Stanbury, J.B. Wyngaarden, D.S. Fredrickson, J.L. Goldstein and M.S. Brown (editors). Mc Graw-Hill Inc. New York, 5th edition (1983), pp 589-621.

2. Kayden, H.J. and M.G. Traber. Clinical,nutritional and biochemical consequencesof apolipoprotein B deficiency. In, Lipoprotein Deficiency Syndromes. A. Angel and J. Frohlich (editors). Adv. in Exper. Med. 201: 67-81, 1986, Plenum Press, New York.

3. Mars, H., Lewis L.A., Robertson, A.L., Butkus, A. and G.M. Williams. Familial hypobetalipoproteinemia a genetic disorder of lipid metabolism with nervous system involvement.Am. J. Med. 46: 886-900, 1969.

4. Cottrill, C., Glueck, C.J., Leuba, V., Millett, F., Puppione, D. and W. V. Brown. Familial homozygous hypobetalipoproteinemia. Metabolism, 23: 779-791, 1974.

5. Polonovski,C., Navarro J., Fontaine, J.L.,.de Guyon, F., Saudubray, J.M. and L. Cathelineau. Maladie d'Anderson. Ann. Pediatr. (Paris) 17: 342-354, 1970.

6. Malloy, M.J., Kane, J.P., Hardman, A., Hamilton, R.N., and K.B. Dalal. Normotriglyceridemic abetalipoproteinemia. Absence of the B100 apolipo-protein. J. Clin. Invest. 67: 1441-1450, 1981.

7. Takashima, Y., Kodama, T., Uda, H., Kawamura, M., Aburtani, H., Itakura, H., Akanuma, Y., Takaku, F. and M. Kawade. Normotriglyceridemic apobetalipoproteinemia in infancy: an isolated apolipoprotein B100 deficiency. Pediatrics, 75: 541-546, 1985.

8. Herbert, P.N., Hyams, J.S., Bernier, O.N., Berman, M.M., Saritelli, A.L., Lynch K.M., Nichol A.V. and T.M. Forte. Apolipoprotein B deficiency Intestinal steatosis despite apolipoprotein B48 synthesis. J. Clin. Invest. 76: 403-412, 1985.

9. Bouma, M.E., Beucler, I., Aggerbeck, L.P., Infante, R., and J. Schmitz. Hypobetalipoproteinemia with accumulation of an apoprotein B-like protein in intestinal cells, immunoenzymatic and biochemical characterization of seven cases of Anderson's disease. J. Clin. Invest. 78: 398-410, 1986.

10. Roy, C.C., Levy, E., Green, P.H.R., Sniderman, A., Retarte, J., Buts, J.P., Orquin, J., Brochu, P., Weber, A.M., Morin, C., Mareil, Y. and R.J. Deckelbaum. Malabsorption, hypocholesterolemia and fat-filled enterocytes increased intestinal apoprotein B. Chylomicron retention disease. Gastroenterology, 92: 390-399, 1987.

11. Steinberg, D., Grundy,S.M., Mok, H.Y.I., Turner, J.D., Weinstein, D.B., Brown V.W. and J.J. Albers. Metabolic studies in an unusual case of asymptomatic familial hypobetalipoproteinemia with hypoalphalipoproteinemia and fasting chylomicronemia. J. Clin. Invest. 64: 292-301, 1979.

12. Vega, G.L., Von Bergmann, K., Grundy,S.L., Beltz, W., Jahn, C., and C. East. Increased catabolism of VLDL-apolipoprotein B and synthesis of bile acids in a case of hypobetalipoproteinemia. Metabolism, 36: 262-269, 1987.

13. Illingworth, D.R., Connor, W.E., and R.G. Miller. Abetalipoproteinemia: report of two cases and review of therapy. Arch. Neurol. 37: 659-662, 1980.

14. Brown, M.S., Kovanen, P.T. and J.L. Goldstein. Regulation of plasma cholesterol by lipoprotein receptors. Science, 212: 628-635, 1981.

15. Brown, M.S., Branan, P.G., Bohmfalk, M.A., Brunschede, G.Y., Dana, S.E., Helgelson, J. and J.L. Goldstein. Use of mutant fibroblasts in the analysis of regulation of cholesterol metabolism in human cells. J. Cell. Physiol. 85: 425-436, 1975.

16. Ho, Y.K., Faust, J.R., Bilheimer, D.W., Brown, M.S. and J.L. Goldstein. Regulation of cholesterol synthesis by low density lipoproteins in isolated human lymphocytes. J. Exp. Med. 145: 1531-1548, 1977.

17. Reichl, D., Myant, N.B. and S.K. Lloyd. Surface binding and catabolism of low density lipoprotein by circulating lymphocytes from patients with abetalipoproteinemia with observations on sterol synthesis in lymphocytes from one patient. Biochem. Biophys. Acta. 530: 124-131, 1978.

18. Illingworth, D.R., Connors, W.E., Buist, W.R., Jhavery,M.B., Lin, D.S. and P.P. McMurry. Sterol balance in abetalipoproteinemia : studies in a patient with homozygous familial hypobetalipoproteinemia. Metabolism, 28: 1152-1160, 1979.

19. Illingworth, D.R., Connors, W.E. and J. Diliberti. Lipid metabolism in abetalipoproteinemia : a study of cholesterol absorption and sterol balance in two patients. Gastroenterology, 78: 68-75, 1980.

20. Myant, N.B., Reichl, D. and J.K. Llyod. Sterol balance in a patient with abetalipoproteinemia. Atherosclerosis, 29: 509-512, 1978.

21. Blum, C.B., Deckelbaum, R.J., Witte, L.D., Tall, A.R. and J. Cornicelli. Role of apolipoprotein E-containing lipoproteins in abetalipoproteir ~ia. J. Clin. Invest. 70: 1157-1169, 1982.

22. Illingworth, D.R., Alam, N.A., Sandberg, E.E., Hagemenas, F.L. and D.L. Layman. Regulation of low density lipoprotein receptors by plasma lipoproteins from patients with abetalipoproteinemia. Proc. Nat. Acad. Sci. USA, 80: 3475-3479, 1983.

23. Kayden, M.J. Abetalipoproteinemia : abnormalities of serum lipoproteins. In, Protides of biological Fluids, Peeters H. (editor). Oxford, Pergamon Press, (1978), pp 271-276.

24. Illingworth, D.R., Corbin, D.R., Kemp, E.D. and E.J. Keenan. Hormones changes during the menstrual cycle in abetalipoproteinemia : reduced luteal phase progesterone in a patient with homozygous hypobetalipoproteinemia. Proc. Nat. Acad. Sci. USA, 79: 6685-6689, 1982.

25. Illingworth, D.R., Kenny, T.A. and E.S. Orwoll. Corticosteroid production in abetalipoproteinemia : evidence for an impaired response to ACTH. J. Lab. Clin. Med. 100: 115-126, 1982.

26. Illingworth, D.R., Kenny, T.A. and E.S. Orwoll. Adrenal function in heterozygous and homozygous hypobetalipoproteinemia. J. Clin. Endocrinol. Metab. 54: 27-33, 1982.

27. Gwynne, J.T. and J.F. Strauss. The role of lipoproteins in steroidogenesis and cholesterol metabolism in steroidogenic glands. Endocr. Rev. 3: 299-329; 1982.

28. Levy, E., Marcel, Y., Deckelbaum, R., Milne, R., Lepage, G., Seidman, E., Bendayan, M. and C. Roy. Intestinal apo-B synthesis,lipids and lipoproteins in chylomicron retention disease. J. Lipid. Res. 28: 1263-1274, 1987.

29. Young, S.G., Bertics, S.J., Curtiss, L.K. and J. Witztum. Characterization of an abnormal species of apolipoprotein B, apolipoprotein B37, associated with familial hypobetalipoproteinemia. J. Clin. Invest. 79: 1831-1841, 1987.

30. Glickman, R.M., Green, P.H.R. and R. Lees. Immunofluorescence studies of apolipoprotein B in normal and abetalipoproteinemia intestinal mucosa. Gastroenterology, 76: 288-292, 1979.

31. Levy, E., Marcel, Y.L., Milne, R.W., Grey, V.G. and C. Roy. Absence of intestinal synthesis of apolipoproein B48 in two cases of abetalipoproteinemia. Gastroenterology, 93: 1119- , 1987.

32. Lackner, K.J., Monge, J.C., Gregg, R.E., Hoeg, J.L., Triche, T.J., Law, S.W. and H.B. Brewer. Analysis of the apolipoprotein B gene and .messenger ribonucleic acid in abetalipoproteinemia. J. Clin. Invest. 78: 1707-1712, 1986.

33. Dullaart, R.P.F., Speelberg, B., Schurman, H.J. Milne, R.W., Havekes, L., Marcel, Y.L., Geuze, H.J., Hulshof, M.M. and D.W. Erkelens. Epitopes of apoliprotein B100 and B48 in both liver and intestine. Expression and evidence for local synthesis in recessive abetalipoproteinemia. J. Clin. Invest. 78:1397-1401, 1986.

34. Li, W.H., Tanimura, M., Luo, C.C., Datta, S. and L. Chan. The apolipoprotein multigene family : biosynthesis, structure-function relationships, and evolution. J. Lipid. Res. 29: 245-265, 1988.

35. Breslow, J.L. Apolipoprotein genetic variation and human disease.
Physiol. Rev. 68: 85-124, 1988.

36. Powell, L.M., Wallis, S.C., Pease, R.J., Edwards, H.Y., Knott, T.J.
and J. Scott. A novel form of tissue specific RNA processing produces
apolipoprotein B48 in intestine. Cell, 50: 831-840, 1987.

37. Chen, S.H., Habib, G., Yang, C.Y., Gu, Z.W., Lee, B.R., Weng, S.A.,
Silberman, S.R., Cai, S.J., Deslypere, J.P., Rosseneu, M., Gotto, A.M.
and L. Chan. Apolipoprotein B48 is the product of a messenger RNA with
an organ specific inframe stop codon. Science, 238: 363-366, 1987.

38. Higuchi, K., Hospattankar, A.V., Law, S.W., Meglin, G.N., Cortright,
J. and H.B. Brewer. Human apolipoprotein B (apo-B) mRNA : identification
of two distinct apo-B mRNAs, an mRNA with the apo-B100 sequence and an
apo-B mRNA containing a premature in frame stop codon, in both liver
intestine. Proc. Nat. Acad. Sci. USA, 85: 1772-1776, 1988.

39. Ross, R.S., Gregg, R.E., Law, S.W., Monge, J.C., Grant, S.M.,
Higuchi, K., Triche, T.J., Jefferson, J. and H.B. Brewer. Homozygous
hypobetalipoproteinemia: a disease distinct from abetalipoproteinemia at
the molecular level. J. Clin. Invest. 81: 590-595, 1988.

40. Talmud, P., Lloyd, J.K., Muller, D.P.R., Collins, D.R., Scott, J.,
and S. Humphries. Genetic evidence from two families that the apolipoprotein B gene is not involved in abetalipoproteinemia. J. Clin. Invest.
82: 1803-1806, 1988.

41. Leppert, M., Breslow, J.L., Wu, L., Hasstedt, S., O'Connell, P.,
Lathrop, M., Williams, R.R., White, R. and J.M. Lalouel. Inference of
a molecular defect of apolipoprotein B in hypobetalipoproteinemia by
linkage analysis in a large kindred. J. Clin. Invest. 82: 847-851, 1988.

42. Young, S.G., Northey, S.T. and B.J. MacCarthy. Low plasma cholesterol levels caused by a short deletion in the apolipoprotein B gene.
Science, 241: 591-593, 1988.

43. Collins, D.R., Knott, T.J., Pease, R.J., Powell, L.M., Wallis, S.C.,
Robertson, S., Pullinger, C.R., Milne, R.W., Marcel, Y.L., Humphries, S.E.
Talmud, P.J., Lloyd, J.K., Miller, N.E., Muller, D. and J. Scott.
Truncated variants of apolipoprotein B cause hypobetalipoproteinemia.
Nucleic Acid Res. 16: 8361-8375, 1988.

THE MOLECULAR BASIS OF ApoC-II DEFICIENCY

Fojo, S.S., *Beisiegel, U., +Stalenhoef, A.F.H., Bojanovski
M., Gregg, R.E., Greten, H., and Brewer, H.B.,Jr.

MDB, NHLBI, NIH, Bethesda, MD; *Medizinische Kern-und
Poliklinik, Universitaets Klinik Eppendorf, Hamburg, FGR;
+St. Radbaud Hospital, Nijmegen, The Netherlands; Medizinische
Hochschule Hannover, Hanover, FGR

SUMMARY

The genetic defect that leads to deficiency of apoC-II in 2 patients
with apoC-II deficiency from independent kindreds has been defined.

The first patient from Hamburg, West Germany, was found to have a single
base substitution (guanosine for a cytosine,) at the first base of the donor
splice site of intron II of the apoC-II gene by sequence analysis. This
mutation should abolish normal splicing at this site and ultimately lead to
the deficiency of plasma apoC-II observed in this kindred. Analysis of total
RNA isolated from the patients liver by Northern, slot blot, and in situ RNA
hybridization, as well as evaluation of the intrahepatic apoC-II content by
immunohistochemistry confirmed the results of our sequence analysis.

The second patient from Nijmegen, the Netherlands was found to have a
deletion of a cytosine at position # 2943 of exon 3 of the apoC-II gene.
This mutation results in the formation of a premature termination codon at a
position corresponding to amino acid #18 of normal apoC-II. As a result,
this mutation leads to the formation of a truncated apoC-II that is unlikely
to activate lipoprotein lipase and thus, leads to apoC-II deficiency in this
kindred.

The characterization of the genetic abnormalities in the apoC-II gene of
these 2 patients will further our understanding of the molecular defects that
can lead to deficiency of apoC-II. These experiments also illustrate the
heterogeneity of lesions that can lead to the syndrome of apoC-II deficiency.

INTRODUCTION

Apolipoprotein (apo) C-II plays a central role in triglyceride metabolism as a cofactor for the enzyme lipoprotein lipase. The importance of
apoC-II as an activator of this enzyme has been established by the absence of
lipoprotein lipase activity in patients with a deficiency of apoC-II. This
syndrome, which is inherited as an autosomal recessive trait, has been
described in several kindreds (1-9). Patients homozygous for apoC-II
deficiency have marked derangements in triglyceride metabolism which include
an elevation of plasma triglycerides, chylomicrons and VLDL, as well as
eruptive xanthomas and an increased incidence of pancreatitis.

The diagnosis of apoC-II deficiency is established by finding a virtual absence of apoC-II in plasma associated with reduced post-heparin lipoprotein lipase activity which is corrected by the addition of normal apoC-II containing plasma. Patients with near normal levels of a nonfunctional apoC-II variant are also designated as patients with apoC-II deficiency. Transient normalization of triglyceride and lipoprotein abnormalities present in the plasma of apoC-II deficiency patients can be achieved by infusion of normal plasma (1,3,7), purified apoC-II (5) or synthetic apoC-II fragments (7).

Of the various kindreds described with a deficiency of apoC-II, two have been identified with an abnormal plasma apoC-II. Both apoC-II$_{Toronto}$, and apoC-II$_{St.Michael}$ have an altered carboxyl-terminal amino acid sequence when compared to normal apoC-II (8,9) that results in the formation of a non-functional protein. These amino acid sequence differences are consistent with a base deletion for apoC-II$_{Toronto}$ and a base insertion for apoC-II$_{St.Michael}$ resulting in a subsequent shift of the translational reading-frame in the apoC-II gene of both patients. Studies of the apoC-II gene in apoC-II deficient patients at the DNA level have been limited to Southern blot hybridization analyses which have not, thus far, revealed any major rearrangements of the apoC-II gene (10,11).

In the present study we investigate the molecular defect at the gene level in 2 kindreds with apoC-II deficiency. A donor splice site mutation in the second intron of the apoC-II$_{Hamburg}$ gene and a deletion in the third exon of apoC-II$_{Nijmegen}$ gene have been identified by sequence analysis of the probands DNA. We propose that these mutations are the basis for the deficiency of apoC-II in these two different kindreds.

MATERIALS AND METHODS

Clinical data - The clinical features of the apoC-II deficient probands studied have been previously described (12,13).

Two dimensional (2) gel electrophoresis: Two-D gel electrophoresis of plasma or VLDL was done by performing isoelectric focussing in the first dimension followed by SDS-polyacrylamide gel electrophoresis in the second dimension (14). The gels were stained for protein detection by the silver stain method or used for protein transfer to nitrocellulose paper (15,16,17). Immunoblotting was done by the indirect immunoperoxidase method as described by the manufacturer's (BioRad, Inc., Richmond, CA) instructions.

Immunohistochemistry: Immunohistochemical analysis of 8 um thick liver sections was done by the immunocolloidal gold method followed by silver enhancement as described by the manufacturer (Auro Probe^{TM} LM, Jansen Life Sciences Products, Olen, Belguim). Mouse monoclonal anti-apoC-II antibody was used as the first antibody.

RNA Studies: Total RNA from normal and two apoC-II deficient patient's liver was isolated by the guanidinium thiocyanate method (18). Hybridization following Northern and slot blot analysis was performed by utilizing a nick-translated probe isolated from an apoC-II cDNA and an apoA-I, β-actin and albumin cDNA clones as controls (19,20). Densitometric scanning of the autoradiogram was then performed. In situ RNA hybridization was performed by hybridizing liver sections with radiolabelled apoC-II sense and antisense strands as described (21).

Southern blot hybridization: Genomic DNA from 2 patients with apoC-II deficiency was isolated from WBC and digested with restriction endonuclease enzymes BamH1, BglI, EcoRI, Hind III, PstI, SphI, SstI. The restriction enzyme fragments were run on 0.7% slab agarose gel electrophoresis and transfered to nitrocellulose paper for Southern blotting. Hybridization was performed by utilizing a 350 bp apoC-II cDNA probe as described (10).

Genomic Libraries: WBC genomic DNA from 2 patients with apoC-II deficiency was partially digested with MboII and used to prepare a genomic library by infecting E.coli P2492 cells with EMBL-3 recombinant phage as described (22). The library was screened for apoC-II cDNA clones by using an apoC-II cDNA probe. DNA was prepared from the isolated clones and sequenced by the dideoxynucleotide chain termination method of Sanger (23).

Sequence amplification with Taq DNA polymerase: One ug of genomic DNA from control and apoC-II deficient subjects was amplified by the automated PCR technique (24) for 30 cycles using TaqI DNA polymerase and 20 base pair primers. The amplified region included base number 2645 through 3102 of the apoC-II gene (22). Two minute extensions at 72°C and 30 second denaturations at 94°C, and primer annealing at 55°C were performed. One tenth of the total amplified DNA was digested with two units of either DdeI or HphI for 2 hrs. The digested fragments were separated on a 4% agarose TAE minigel at 85 volts for 1 hour. DNA was identified by staining with ethidium bromide.

A Two Dimensional
Electrophoresis

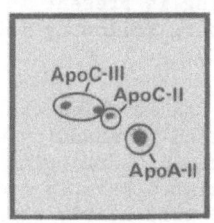

NORMAL PLASMA

B Two Dimensional
Electrophoresis

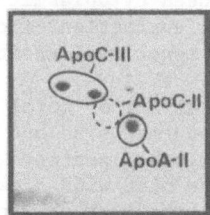

PATIENT PLASMA

C   Immunoblot

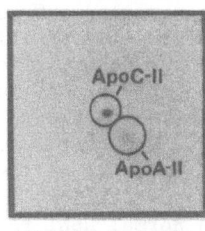

NORMAL PLASMA

D   Immunoblot

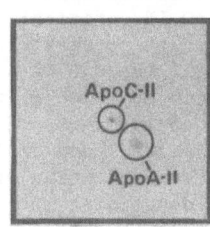

PATIENT PLASMA

Fig. 1   Two dimensional gel electrophoretograms and immunoblots of plasma from a normal subject and the Hamburg proband. Silver stain of plasma proteins from normal (A) and patient (B), as well as immunoblot analysis of plasma from normal (C) and patient (D) are illustrated. The normal location of apoC-II in the electrophoretogram (B) is indicated by the dotted circle. ApoA-II standard has been included in panels C and D as reference.

Analysis of plasma apoC-II in the proband from Hamburg by 2.0 gel electrophoresis and immunoblotting is illustrated in Figure 1. Silver staining did not detect any apoC-II in the patient's plasma (Figure 1B) but the more sensitive method of immunoblotting detected reduced amounts of a normally migrating apoC-II protein (Figure 1D).

Immunohistochemical analysis of liver incubated with an anti-apoC-II monoclonal antibody revealed low but detectable intrahepatic levels of apoC-II in the patient (Figure 2C) when compared to normal (Figure 2B) based on the intensity of the brown staining. Specificity of the procedure is demonstrated by the lack of staining within connective tissue which is not expected to contain C-II apolipoprotein (Figure 2B).

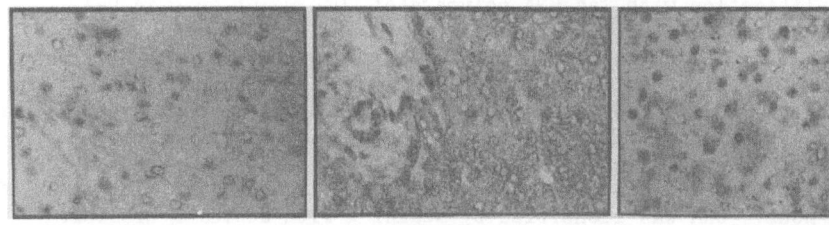

| A. Normal Liver | B. Normal Liver | C. Patient Liver |
| Irrelevant Antibody | Anti apo C-II | Anti apo C-II |

Fig. 2   Immunohistochemical analysis of normal liver incubated with an irrel-
event antibody (A) and an anti-apoC-II monoclonal antibody (B).
Markedly reduced cytoplasmic staining is present in the liver of the
apoC-II deficient patient from Hamburg following incubation with the
anti-apoC-II monoclonal antibody (C).

Northern blot analysis of total liver RNA revealed the apoC-II mRNA of the patient to be of normal size but decreased in quantity when compared to the apoC-II mRNA from other patients (Figure 3). Stripping and rehybridization of the blot with nick translated B-actin and apoA-I cDNA probes followed by autoradiography revealed that nearly equal levels of both of these mRNAs were present in hepatocytes from both the apoC-II deficient patient and normal subjects (Figure 3).

Sequence analysis of two apoC-II genomic clones isolated from the DNA library of the apoC-II deficient patient revealed a G to C substitution at the highly conserved first nucleotide position in the 5' splice site of intron II (Figure 4A). Consensus donor splice site and normal apoC-II sequences are also illustrated in Fig. 4A. Autoradiographs of sequence gels from normal and apoC-II deficient subjects illustrating the G to C substitution are shown in Fig. 4C. Computer analysis (Fristensky- Cornell DNA Sequencing Analysis Program) of the patient's apoC-II gene sequence containing the G to C substitution, revealed a new DdeI site at position 2715 and the loss of a normal HphI restriction site at position 2725 as compared to the normal apoC-II gene (22).

Amplification of the apoC-II gene from normal and apoC-II deficient subjects was performed using the polymerase chain reaction and synthetic oligonucleotides illustrated in Figure 5A. The amplified product was 457 bp in length. Digestion of DNA amplified from a normal subject with DdeI generated two fragments of 393 and 64 bp in size but digestion of the patients amplified DNA with the same enzyme resulted in the formation of three fragments of 69, 324 and 64 bp in size.

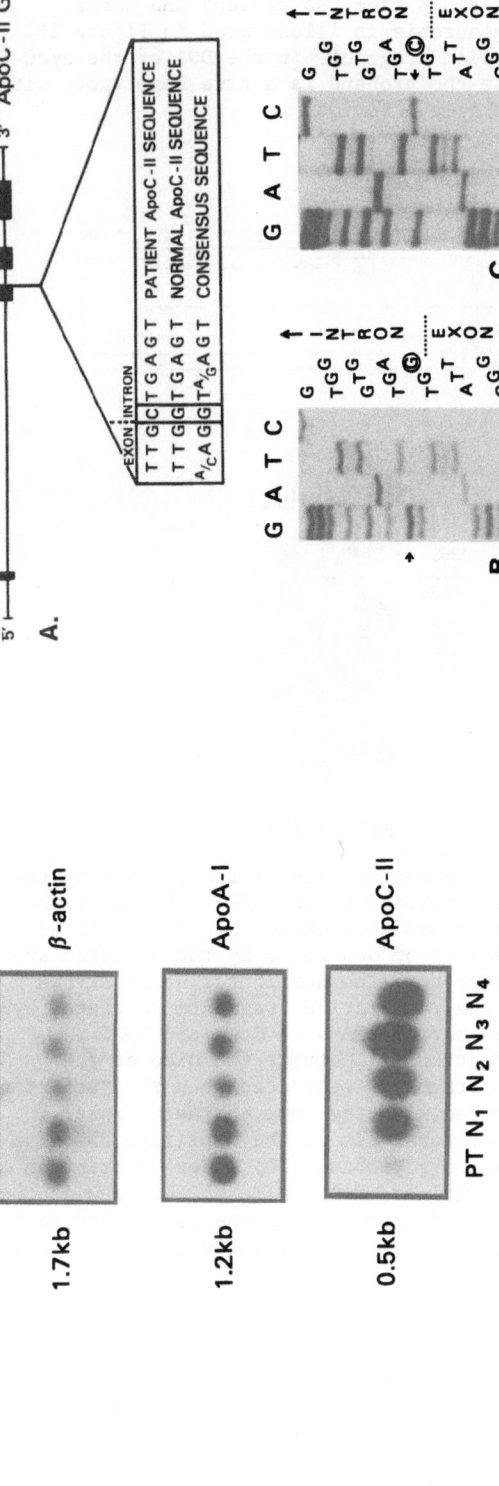

Fig. 3   Panel A is a schematic representation of
the normal apoC-II gene. Exons are illustrated
by the solid bars interrupted by lines that
represent introns. The patient, normal and
consensus donor splice site sequences for
intron II are shown. The G to C mutation is
highlighted by a box. Panels B and C contain
the autoradiographs of sequencing gels of DNA
from normal and apoC-II deficient subjects near
the region of the mutation. The G to C substitution
is indicated by arrows.

Fig 4.   Northern blot analysis of total liver RNA
from normal subjects (N) and the Hamburg patient
(PT) hybridized with β-actin, apoA-I, and Apoc-II
cDNA probes.

Digestion of DNA amplified from a normal subject with HphI generated four fragments of 80, 202, 28 and 147 bp in size while digestion of the patient's amplified DNA with the same enzyme resulted in the formation of three fragments of 282, 28, and 147 bp in length. Analysis of the restriction enzyme fragments of the amplified DNA from the patient and normal subjects by agarose gel electrophoresis is illustrated in Figure 5B. Absence of normal sized restriction fragments in the DNA of the apoC-II deficient patient confirmed that the proband is a true homozygote with both apoC-II alleles having the same mutation.

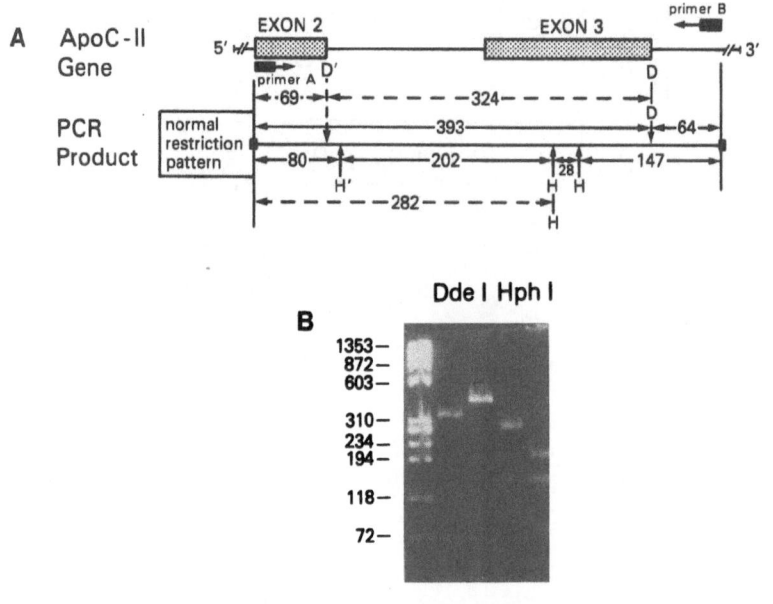

Fig. 5   Panel A is a schematic representation of the portion of the apoC-II gene that was amplified by the PCR. The positions of the primers (A and B) are indicated. New DdeI (D') and HphI (HJ') sites generated or destroyed by the mutation are indicated by the apostrophe. The horizontal solid arrows indicate the size of the restriction fragments generated by digestion of amplified normal DNA with DdeI and HphI. The horizontal dashed arrows illustrate the size of the different restriction fragments generated when the patient's amplified DNA is digested with the same enzymes. Panel B shows an electrophoretic analysis of the amplified DNA from normals (N) and the Hamburg proband (PT) after restriction enzyme digestion with DdeI and HphI. The differences in the sizes of the restriction fragments from the normal and patient DNA are illustrated. DNA molecular weight markers are shown in the lane designated M.

Figure 6 illustrates the 2-dimensional gel electrophoretograms of proteins present in the plasma of a normal subject (Figure 6A) and the apoC-II deficient proband from Nijmegen (Figure 6B). No apoC-II was detected in the patient's plasma by either silver staining (Fig. 6B) or immunoblot analysis (Figure 6D). ApoC-II was easily detected in normal plasma (Figures 6A,C).

Sequence analysis of an apoC-II genomic clone isolated from the DNA library of the apoC-II deficient patient revealed a deletion of a guanosine at position 2943 (Fig. 7A) within exon 3 of the normal apoC-II gene (22). Gels shown demonstrate the sequence of the complementary DNA strand. Figure 7B illustrates the autoradiograms of sequencing gels of the complementary DNA strand from both normal and apoC-II deficient subjects. The position of the base deletion in the patient's gene is indicated by the arrow. This mutation results in the introduction of a premature termination codon (TGA) at this position of the apoC-II gene.

Computer analysis (Fristensky-Cornell DNA Sequencing Analysis Program) of the apoC-II sequence from the apoC-II deficient subject containing the guanosine deletion revealed the loss of a normal HphI restriction site at base 2954 as compared to the normal apoC-II gene(22). The apoC-II gene from normal and apoC-II deficient subjects were amplified by the polymerase chain reaction (Fig. 8A). The amplified product was 457 bp in length. Digestion of the amplified normal DNA with HphI is expected to generate 4 fragments of 80, 202, 28 and 147 bp in length, but digestion with HphI of the amplified patient's DNA containing the guanosine deletion, resulted in the formation of 3 fragments of 80, 202 and 174 bp in length (Fig. 8A). Agarose gel electrophoresis of the restriction enzyme fragments generated after digestion of DNAs from normal and apoC-II deficient subjects with HphI is illustrated in Fig. 8B. Total absence of the normal sized restriction fragment (147 bp) and presence of only the abnormal sized fragment (174 bp) established that the proband is a homozygote with both apoC-II alleles containing the guanosine deletion.

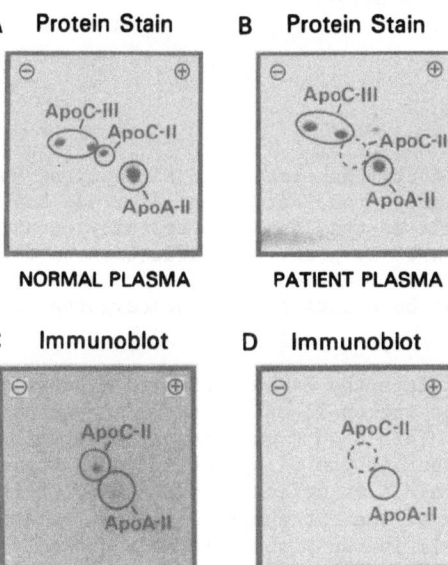

Fig. 6   Two-dimensional gel electrophoretogram of plasma apolipoproteins from a normal subject and the Nijmegen proband. The dashed circles indicate the position of normal apoC-II not identified in the proband's plasma by protein stain (B) or immunoblotting (D). ApoA-II standard has been included in all panels as reference.

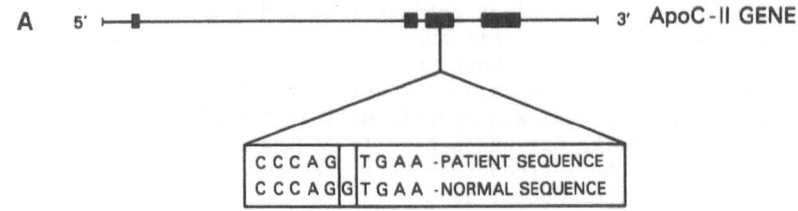

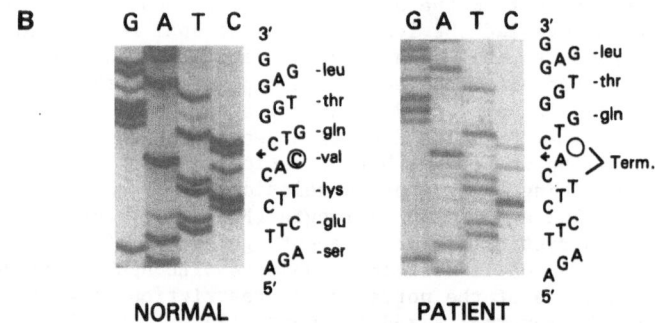

Fig. 7 Panel A contains a schematic representation of the apoC-II gene containing the position of the guanosine deletion identified in the patient from Nijmegen. Autoradiograms of sequencing gels from the normal and apoC-II deficient subjects in the region of the mutation are illustrated in panel B. The sequence shown is complementary to the sense strand of the apoC-II gene. The guanosine deletion is indicated by an arrow.

DISCUSSION

The genetic defects that lead to deficiency of apoC-II in 2 patients with apoC-II deficiency from independent kindreds have been elucidated. Both mutations result in the loss of a normally occurring HphI restriction enzyme site in the apoC-II gene. Amplification of the mutant DNA sequence by the polymerase chain reaction and restriction enzyme digestion with HphI has established that both patients are homozygotes for their respective defect.

In the patient from the Hamburg kindred a substitution of a G for a C at the first base of intron II of the apoC-II gene has been identified by sequence analysis. This G to C mutation alters the highly conserved dinucleotide invariant at the 5' splice sites in structural genes and would be anticipated to result in defective processing of the apoC-II$_{Hamburg}$ mRNA (25, 26). Analysis of the intrahepatic content of apoC-II mRNA and protein and of plasma apoC-II in the proband reveals markedly reduced but detectable amounts of both. Thus, despite the mutation of the critical GT dinucleotide of the apoC-II$_{Hamburg}$ gene, a small amount of correctly spliced message is made resulting in the low but detectable levels of normal sized apoC-II message visualized by our Northern studies. The amount of normal apoC-II in the patients' plasma however, is not sufficient to activate lipoprotein lipose. Thus, in this kindred deficiency of apoC-II is a result of a G to C substitution within the donor splice site of the second intron of the apoC-II$_{Hamburg}$ gene that leads to abnormal splicing and markedly reduced plasma apoC-II levels.

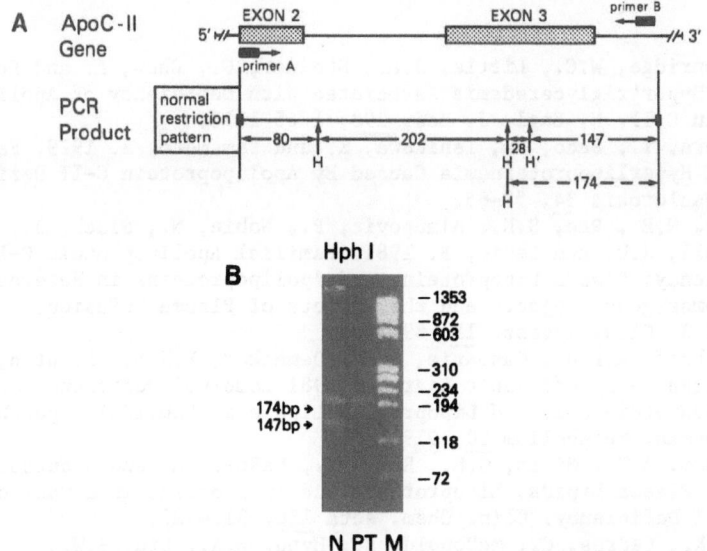

Fig. 8 A schematic representation of the portion of the apoC-II gene amplified by the PCR reaction using primers A and B is shown in panel A.
Normal HphI sites are indicated. The HphI site eliminated by the
mutation is indicated as H'. The horizontal solid arrows indicate
the size of the restriction fragments generated by digestion of
amplified normal DNA with HphI. The horizontal dashed arrows
illustrate the size of the unique restriction fragment generated
when the DNA from the Nijmegen patient is digested with the same
enzyme. Panel B contains an electrophoretic analysis of the amplified DNA from normal (N) and the patient (PT) after restriction
enzyme digestion with HphI. The differences in the size of the
restriction fragments from the normal and patient DNA are illustrated. DNA molecular weight markers are shown in the lane designated M.

In the patient from the Nijmegen kindred sequence analysis studies have
revealed a deletion of base 2943 of the third exon of the apoC-II gene. This
mutation results in a shift of the reading frame and introduction of an in
frame premature termination codon that leads to the synthesis of a truncated
17-amino acid apoC-II protein. This abnormal protein, if secreted normally,
is unlikely to function as an activator of lipoprotein lipase. Thus, the
primary genetic defect in the Nijmegen kindred is the deletion of a guanosine
in the third exon of the apoC-II_Nijmegen gene which results in the formation
of a truncated apoC-II protein and a deficiency of functional apoC-II.

These studies will help us to better understand the molecular defects
that can lead to apoC-II deficiency and illustrate the genetic heterogeneity
that underlies this syndrome.

REFERENCES

1. Breckenridge, W.C., Little, J.A., Steiner, G., Chow, A. and Poapst, M. 1978. Hypertriglyceredemia Associated With Deficiency of Apolipoprotein C-II. N. Engl. J. Med. 298, 1265-1273.

2. Yamamura, T., Sudo, H., Ishikawa, K. and Yamamoto, A. 1979. Familial Type I Hyperlipoproteinemia Caused By Apolipoprotein C-II Deficiency. Atherosclerosis 34, 53-65.

3. Miller, N.E., Rao, S.N., Alaupovic, P., Noble, N., Slack, J., Brunzell, J.D. and Lewis, B. 1981. Familial Apolipoprotein C-II Deficiency: Plasma Lipoproteins and Apolipoproteins in Heterozygous and Homozygous Subjects and the Effects of Plasma Infusion. Europ. J. Clin. Invest. 11, 69-76.

4. Stalenhoef, A.F.H., Casparie, A.F., Demacker, P.N.M., Stouten, J.T.J., Lutterman, J.A. and van't Laar, A. 1981.Combined Deficiency of Apolipoprotein C-II and Lipoprotein Lipase in Familial Hyperchylomicronemia. Metabolism 30, 919-926.

5. Catapano, A.I., Mills, G.L., Roma, P., LaRosa, M. and Capurso, A. 1983. Plasma Lipids, Lipoproteins and Apoproteins in a Case of apoC-II Deficiency. Clin. Chem. Acta 130, 317-327.

6. Saku, K., Cedres, C., McDonald, B., Hynd, B.A., Liu, B.W., Srivastava, L.S., and Kashyap, M.L. 1984. C-II Anapolipoprotenemia and Severe Hypertriglyceridemia. The American Journal of Medicine 77, 457-462.

7. Baggio, G., Manzato, E., Gabelli, C., Fellin, R., Martini, S., Baldo Enzi, G., Verlato, F., Baiocchi, M.R., Sprecher, D.L., Kashyap, M.L., Brewer, H.B., Jr. and Crepaldi, G. 1986. Apolipoprotein C-II Deficiency Syndrome. J. Clin. Invest. 77, 520-527.

8. Connelly, P.W., Maguire, G.F., Hofmann, T. and Little, J.A. 1987. Structure of Apolipoprotein C-II$_{Toronto}$, a Nonfunctional Human Apolipoprotein. Proc. Natl. Acad. Sci. USA 84, 270-273.

9. Connelly, P.W., Maguire, G.F. and Little, A.J. 1987. Apolipoprotein C-II$_{St. Michael}$; Familial Apolipoprotein C-II Deficiency Associated With Premature Vascular Disease. J. Clin. Invest., 80, 1597-1606.

10. Fojo, S.S., Law, S.W., Sprecher, D.L., Baggio, G. and Brewer, H.B., Jr. 1984. Analysis of the apoC-II Gene in apoC-II Deficient Patients. Biochem. Biophys. Res. Com. 124, 308-313.

11. Humphries, S.E., Williams, L., Myklebost, O, Statenhoef, A.F., Demacker, P.N.M., Baggio, G., Crepaldi, G., Galton, D.J., and Williamson, R. 1984. Familial Apolipoprotein C-II Deficiency: A Preliminary Analysis of the Gene Defect in Two Independent Families. Hum. Genet. 67, 151-155.

12. Stalenhoef, A.F.H., Casparie, A.F., Demalker, P.N.M., Stousen, J.T.J., Lutterman, J.A., and Van't Laar, A. (1981)

13. Fojo, S.S., Beisiegel, U., Beil, U., Higuchi, K., Bojanovski, M., Gregg, R.E., and Brewer, H.B.,Jr. A Donor Splice Site Mutation in the apoC-II gene (apoC-II$_{Hamburg}$) in a Patient With apoC-II Deficiency. JCI (in press)

14. Sprecher, D.L., Taam, L., and Brewer, H.B., Jr.1984. Two Dimensional Electrophoresis of Human Plasma Apolipoproteins. Clin. Chem. 30, 2084-2092.

15. Merril, C.R., Goldman, D., Sedman, S.A., and Ebert, M.H. 1981. Ultrasensitive Stain for Proteins in Polyacrylamide Gels Show Regional Variation in Cerebrospinal Fluid Proteins. Science 211, 1437-1438.

16. Morrisey, J.H. 1981. Silver Stain for Proteins in Polyacrylamide gels: A Modified Procedure With Enhanced Uniform Sensitivity. Anal. Biochem 117, 307-310.

17. Towbin, H., Staehlin, T., and Gordon, T. 1979. Electrophoretic Transfer of Proteins from Polyacrylamide Gels to Nitrocellulose Sheets - Procedure and Some Applications. Proc. Natl. Acad. Sci. USA 76, 4350-4354.17.

18. Chirgwin, J.A., Przybyla, A.E., MacDonald, R.J. and Rutter, W.J. 1979. Isolation of Biologically Active Ribonucleic Acid From Sources Enriched in Ribonuclease. Biochemistry 18, 5294-5299.

19. Law, S.W. and Brewer, H.B., Jr., 1984. Nucleotide Sequence and the Encoded Ammoacids of Human Apolipoprotein A-I mRNA. Proc. Natl. Acad. Sci. USA 81, 66-70.

20. Ponte, P., Ng, S.-Y., Engel, J., Gunning, P. and Kedes, L. 1984. Evolutionary Conservation in the Untranslated Regions of Actin mRNAs: DNA Sequence of a Human Beta-Actin. Nucleic Acid. Res. 12, 1687-1696.

21. Harper, M.E., Marselle, L.M., Gallo, R.C. and Wong-Staal, F. 1986. Detection of Lymphocytes Expressing Human T-Lymphotropic Virus Type III in Lymph Nodes and Pheripheral Blood From Infected Individuals Lying in Situ Hybridization. Proc. Natl. Acad. Sci, U.S.A. 83, 772-776.

22. Fojo, S.S., Law, S.W., and Brewer, H.B.,Jr. 1987. The Human Preproapolipoprotein C-II gene: Complete Nucleic Acid Sequence and Genomic Organization. FEBS LETTERS, 213, 221-226.

23. Sanger, F., Coulson, A.R., Barell, B.G., Smith, A.J.H., and Roe, B.A. 1980. Cloning in Single-Stranded Bacteriophage as an Aid to Rapid DNA Sequencing. J. Mol. Biol. 143, 161-178.

24. Wong, C., Dowling, C.E., Saiki, R.K., Higuchi, R.G., Erlich, H.A., Kazazian, H.H., Jr. 1987. Characterization of B-Thalassaemia Mutations Using Direct Genomic Sequencing of Amplified Single Copy DNA. Nature, 330, 384-386.

25. Breathnach, R., and Chambon, P. 1981. Organization and Expression of Eukaryotic Split Genes Coding for Proteins. Ann. Rev. Biochem. 50, 349-383.

26. Mount, S.M. 1982. A Catalogue of Splice Junction Sequences. Nucleic Acid Res. 10, 459-472.

FAMILIAL CHYLOMICRONEMIA DUE TO MUTATIONS IN APOLIPOPROTEIN CII:

APOLIPOPROTEIN CII-TORONTO AND APOLIPOPROTEIN CII-ST. MICHAEL.

Philip W. Connelly, Graham F. Maguire and J. Alick Little

Departments of Medicine and Biochemistry, University of
Toronto and Lipid Research Clinic, St. Michael's Hospital
Toronto, Canada

INTRODUCTION

In this article, we present a summary of our work in establishing the molecular defect in two separate families that were characterized as having familial chylomicronemia due to a deficiency of apolipoprotein CII (apoCII).

ApoCII is important for the normal metabolism of plasma triglyceride by virtue of it's function as an activator of lipoprotein lipase (LPL). ApoCII activates LPL 5 to 10 fold in vitro. The discovery of apoCII deficiency by Breckenridge and colleagues [1] was important because it established the physiological significance of apoCII-dependent activation of LPL. The first demonstration of apoCII deficiency resulted from the investigation of a male chylomicronemic patient who had been identified as having a deficiency of post-heparin LPL. However, upon receiving an infusion of plasma, the plasma triglyceride levels dropped rapidly, and gradually recovered to preinfusion levels several days after the infusion. This indicated that plasma contained a factor, apoCII, that could normalize the plasma triglycerides in this patient. Breckenridge and colleagues went on to identify 14 homozygotes and 28 obligate heterozygotes with apoCII deficiency in this original kindred [2]. Subsequently a number of unrelated cases of familial chylomicronemia due to a deficiency of apoCII have been described [3].

APOCII-TORONTO

Further study of the apoproteins of the homozygotes and heterozygotes from the family first reported to have apoCII deficiency resulted in the discovery of two isoforms of an abnormal protein that reacted with antiserum to human apoCII. Using isoelectric focusing and immunoblot techniques, the major isoform, with a pI of 5.54 was designated apoCII-Y and the minor isoform was designated as apoCII-X [4]. These were readily distinguished from normal apoCII which has a significantly more acidic pI of 4.88. Two-dimensional electrophoretic analysis with isoelectric focusing in the first dimension and sodium dodecyl sulfate acrylamide gel electrophoresis in the second dimension demonstrated that both isoforms had a molecular weight that was slightly smaller than normal apoCII.

To identify the alteration of apoCII that resulted in these marked differences from normal and the loss of function, the major isoform, apoCII-Y, which we have designated as apoCII-Toronto (apoCII-T), was purified from the apoproteins of chylomicrons and VLDL of a homozygote [5]. The isolated VLDL and chylomicrons were treated with acetone to obtain an acetone soluble fraction and then apoCII-T was purified by a single preparative isoelectric focusing step.

The amino acid composition of apoCII-T was compared with the composition of normal apoCII isolated by the identical protocol and the composition derived from the sequence of normal apoCII published by Hospattanker et al. [6]. There were several differences between the composition of apoCII-T and normal apoCII. For instance apoCII-T contains fewer Glx residues than normal apoCII and more Phe residues. Thus it was apparent that the mutant differed from normal apoCII by more than a single amino acid.

The technique that was used to obtain an amino acid analysis (hydrolysis with 6N HCl) cannot be used to measure cysteine or tryptophan. We further investigated the mutant for the presence of a cysteine residue using the reactivity of cysteine with iodoacetate and monitoring the charge of the apoprotein by isoelectric focusing. ApoCII-T shifted to an acidic pI after carboxymethylation with iodoacetate, consistent with the introduction of a single negative charge and the presence of a single cysteine residue in apoCII-T.

We also analyzed the amino acid sequence of intact apoCII-T by automated Edman techniques using a Beckman Model 890C sequenator and found that the N-terminal 38 residues of apoCII-T had a sequence identical to that of normal apoCII as reported by Hospattankar et al. [6]. This included the single tryptophan residue known to be at position 26 in normal apoCII. Thus apoCII-T contained a cysteine residue and a normal N-terminal sequence.

The next step was to generate tryptic peptides of apoCII-T and separate these by reverse-phase HPLC. Normal apoCII contained six tryptic peptides. A total of eight peptides would be predicted on the basis of the sequence of apoCII. From the amino acid analysis of the six peptides it was deduced that the two peptides that were not detected were residues 49 and 50, Leu-Arg and residues 77 to 79, Gly-Glu-Glu.

Five peptides of apoCII-T were identical to normal apoCII, corresponding to residues 1-55 of normal apoCII. However, there was no peptide corresponding to residues 56 through 75 of normal apoCII. Instead, we found a new peptide labeled P6'. To identify the peptide containing the cysteine residue we derivatized apoCII-T with 4-vinyl pyridine, since pyridylethyl-cysteine absorbs strongly at 254 nm. The tryptic peptide profile of pyridylethylated apoCII-T was identical to that of underivatized apoCII-T except for the appearance of a new peptide, identified as P7'. Both P6' and P7' were isolated and analyzed by amino acid composition and sequence techniques. A summary of the results of those analyses is shown schematically in Figure 1.

The sequence of P6' was found to be identical to residues 56 - 69 of normal apoCII. At that point the sequence changed to Thr Lys. This sequence does not occur anywhere in normal apoCII. Inspection of the sequence of the apoCII gene [7] revealed that this change could result from a base deletion in the codon for residue 69, Thr, or residue 70, Gln, and a subsequent translation reading frame shift. The new sequence is predicted to continue beyond Lys as Phe-Phe-Leu-Cys. Amino acid sequence analysis of P7' showed Phe-Phe-Leu-(pyridylethyl)Cys.

Table 1. Lipoprotein Cholesterol and Triglyceride
Concentrations of the ApoCII-S Proband and ApoCII-T Homozygotes

		ApoCII-S Proband mg/dl	ApoCII-T Homozygotes* mg/dl
Plasma	Chol	190	268
	TG	1325	2284
Chylomicrons	Chol	79	160
	TG	745	1813
VLDL	Chol	69	97
	TG	393	532
LDL	Chol	25	18
	TG	51	46
HDL	Chol	17	10
	TG	40	27

* Values for apoCII-T homozygotes are the average for six female
subjects. Abbreviations: Chol, total cholesterol; TG, triglyceride.

Thus apoCII-T differs from normal apoCII starting at position 70.
The significance of this mutation will be considered below.

APOCII-ST. MICHAEL

We began a detailed investigation of another family with hyperchylomicronemia because of a number of unique characteristics [8]. The proband
presented as a 60-year-old female with fasting chylomicronemia. Her
brother also had chylomicronemia and their parents marriage was consanguin-
eous. They also reported a family history of ischemic vascular disease.

Table 1 shows the plasma lipid values of the proband in comparison
with the mean plasma lipid values of seven female homozygotes from the
apoCII-T kindred. It is apparent that the proband had LDL and HDL
cholesterol concentrations below the Lipid Research Clinics' population
fifth percentile and triglyceride-rich LDL and HDL, as well as chylomicronemia and elevated VLDL. This profile is typical for apoCII deficiency.
To determine the basis for the chylomicronemia, post heparin lipases and
the ability of plasma to activate bovine milk lipoprotein lipase were
measured.

The post-heparin plasma lipoprotein lipase activity was found to be
slightly below the normal range, but was not deficient. The post-heparin
plasma hepatic lipase activity was below the normal range, a result
similar to that reported by Breckenridge et al. for apoCII-T homozygotes
[9]. The plasma of the proband activated bovine milk lipase to 1/7th of
the normal level. This is identical to the activation that was seen
using the plasma of apoCII-T homozygotes and is essentially background
for our assay. Thus the chylomicronemia was due to a functional deficiency
of apoCII. To determine whether the deficiency was due to the absence of
apoCII or the presence of a mutant apoCII, the apoproteins of the chylomicron and VLDL lipoproteins were evaluated by isoelectric focusing.

The isoelectric focusing pattern for the apoVLDL of the proband was examined in the presence and absence of the reducing agent dithiothreitol. Under non-reducing conditions, the pattern in the region of the C apoproteins was complex with multiple bands. This pattern resolved under reducing conditions to an apparently normal pattern with a band in the position of normal apoCII. This suggested that a mutant apoCII was present that contained a cysteine residue. This pattern was also altered by either the introduction of a negative charge by carboxymethylation with iodoacetate or introduction of a positive charge by pyridylethylation with 4-vinyl pyridine. Immunoblot confirmed that an apoCII mutant was present and that the pI of this mutant shifted as a result of carboxymethylation or pyridylethylation, consistent with the presence of a single cysteine residue.

The two-dimensional electrophoretic pattern of apoVLDL under reducing conditions showed that the mutant apoCII, designated apoCII-St. Michael (apoCII-S), had a pI indistinguishable from normal apoCII but a molecular weight of 12,300, compared with 8,900 for normal apoCII. To understand the heterogeneity that was observed under non-reducing conditions, the apoVLDL was examined by two-dimensional electrophoresis without reduction.

Spots were identified as apoCII at three different molecular weights and found to correspond to monomeric apoCII-S, apoCII-S:apoCII-S dimers, and apoCII-S:apoA-II dimers. This pattern explained the multiple bands that had been observed in one-dimensional isoelectric focusing under non-reducing conditions. To determine the structural mutation in apoCII-S, pyridylethylated apoCII-S was purified by preparative isoelectric focusing, its amino acid composition determined and its tryptic peptide profile was analyzed.

As seen for apoCII-T, five peptides of apoCII-S were identified that corresponded to residues 1-55 of normal apoCII. However, as with apoCII-T, there was no peak corresponding to residues 56-75 of normal apoCII. There were two new peaks, designated as P6' and P3.1'. These were isolated and their amino acid composition and sequence was determined.

In the sequence of P6', 20 of 23 amino acid residues were positively identified, while in the sequence of P3.1' 11 of 18 residues were positively identified. The major residue of the first cycle of P6', the amino terminus of P6', was identified as Asp, while a minor residue was identified as Ser. The subsequent sequence was identical to that of P6 of normal apoCII, corresponding to residues 57-69 of normal apoCII. After residue 69, the sequence was completely different and did not correspond to any portion of normal apoCII.

Figure 1 is a schematic summary of the results of the amino acid sequence analysis. ApoCII-S was found to be identical to normal apoCII through residue 69. Residue 70, Gln, of normal apoCII was replaced by Pro. Inspection of the DNA sequence of the apoCII gene revealed that this could be explained by the insertion of a single base in the codon for residue 69, Thr, or residue 70, Gln, with a subsequent translation reading frame shift predicting the sequence that was observed for P6' and P3.1'. Thus the lack of function of apoCII-S is due to a structural mutation resulting in the loss of residues 70-79 of normal apoCII and the introduction of a new sequence from residues 70-96.

FUNCTIONAL SIGNIFICANCE OF MUTATIONS IN APOCII

Three functional domains have been postulated in apoCII [10,11], a lipid binding domain between residues 14 and 50, consisting of 3 amphipathic α-helices, a LPL activation domain and a LPL binding domain. The LPL activation domain is thought to consist of residues 55-65, while the LPL binding domain has been postulated to consist of residues 65-75, on the basis of the predicted β-sheet structure of this part of apoCII. The mutations in both apoCII-T and apoCII-S have occurred in the putative LPL binding domain of apoCII. This is consistent with the importance of the carboxy-terminal portion of normal apoCII. It is also possible that the effect of the mutations has been to alter the structure of apoCII within the activation domain.

SUMMARY

In summary, we have described two different forms of apoCII deficiency. Each is due to a mutation in the apoCII gene that has resulted in the production and secretion of a mutant apoCII. Both mutants differ from normal apoCII at nearly the same starting residue, 69 (apoCII-T) and 70 (apoCII-S).

The families differ in that the apoCII-S patients suffer from ischemic vascular disease, while ischemic vascular disease is relatively absent in the apoCII-T patients. Possible candidates as causal agents for the presence of disease in the apoCII-S patients are the presence of apoE4, the presence of apoCII-S or the formation of apoCII-S:apoAII dimers. One must be careful to assign a causative role for the development of disease within the apoCII-S family to the apoE-apoCII gene complex. However, it is important to appreciate that the mechanism for the development of disease within this family is operating within the context of a severe metabolic impairment in the LPL/apoCII dependent hydrolysis of plasma triglycerides.

ApoCII
68   79
Thr Asp Gln Val Leu Ser Val Leu Lys Gly Glu Glu

ApoCII-T
68   74
Thr **Thr Lys Phe Phe Leu Cys**

ApoCII-S
68   79
Thr Asp **Pro Ser Ser Phe Cys Ala Glu Gly Arg Gly**

80   96
Val Thr Ala Arg Pro Pro Ile Ser Gly Gln Gly Glu Ser Pro Leu Leu Pro

Figure 1. The amino acid sequence of the carboxy-terminal residues 56 to 79 of normal apoCII, residues 68 to 74 of apoCII-T and residues 68-96 of apoCII-S. The mutant portions of apoCII-T and apoCII-S are presented in bold type.

REFERENCES

1. Breckenridge, W.C., J.A. Little, G. Steiner, A. Chow, and M. Poapst. 1978. Hypertriglyceridemia associated with deficiency of apolipoprotein CII. N. Engl. J. Med. 298:1265-1273.
2. Cox, D.W., W.C. Breckenridge, and J.A. Little. 1978. Inheritance of apolipoprotein CII deficiency with hypertriglyceridemia and pancreatitis. N. Engl. J. Med. 299:1421-1424.
3. Breckenridge, W.C. 1986. Apolipoprotein CII deficiency. Adv. Exp. Med. Biol. 201:211-226.
4. Maguire, G.F., J.A. Little, G. Kakis, and W.C. Breckenridge. 1984. Apolipoprotein CII deficiency associated with nonfunctional mutant forms of apolipoprotein CII. Can. J. Biochem. Cell Biol. 62:847-852.
5. Connelly, P.W., G.F. Maguire, T. Hofmann, and J.A. Little. 1987. Structure of apolipoprotein CII-Toronto, a nonfunctional human apolipoprotein. Proc. Natl. Acad. Sci. USA. 84:270-273.
6. Hospattankar, A.V., T. Fairwell, R. Ronan, and H.B. Brewer, Jr. 1984. Amino acid sequence of human plasma apolipoprotein CII from normal and hyperlipoproteinemic subjects. J. Biol. Chem. 259:318-322.
7. Sharpe, C.R., A. Sidoli, C.S. Shelley, M.A. Lucero, C.C. Shoulders, and F.E. Baralle. 1984. Human apolipoproteins AI, AII, CII and CIII, cDNA sequences and mRNA abundance. Nucleic Acids Res. 12:3917-3933.
8. Connelly, P.W., G.F. Maguire, and J.A. Little. 1987. Apolipoprotein CII-St. Michael. Familial apolipoprotein CII deficiency associated with premature vascular disease. J. Clin. Invest. 80:1597-1606.
9. Breckenridge, W.C., J.A. Little, P. Alaupovic, D. Cox, F.T. Lindgren, and A. Kuksis. 1983. Abnormal triglyceride clearance secondary to familial apolipoprotein CII deficiency. In The Adipocyte and Obesity: Cellular and Molecular Mechanisms. A. Angel, C.H. Hollenberg, and D.A.K. Roncari, editors. Raven Press, New York. 137-147.
10. Smith, L.C., J.C. Voyta, A.L. Catapano, P.K.J. Kinnunen, A.M. Gotto, Jr., and J.T. Sparrow. 1980. Activation of lipoprotein lipase by synthetic fragments of apolipoprotein CII. Ann. NY Acad. Sci. 348:213-221.
11. Balasubramaniam, A., R.A. Demel, R.F. Murphy, J.T. Sparrow, and R.L. Jackson. 1986. Substitution of $Ser_{61} \rightarrow Gly_{61}$ in human apolipoprotein CII does not alter its activation of lipoprotein lipase. Chem. Phys. Lipids 39:341-346.

APOLIPOPROTEIN C-II DEFICIENCY SYNDROME: NEW INSIGHTS INTO THE MOLECULAR MECHANISM LEADING TO THE DISEASE IN THE APO C-II$_{PADOVA}$ KINDRED

G. Baggio, *S. Fojo, C. Gabelli, S. Martini, L. Previato, C. Corti, G. Crepaldi and *H.B. Brewer Jr.

Department of Internal Medicine, University of Padova, Italy, and *Molecular Disease Branch, NHLBI, NIH, Bethesda, MD, U.S.A.

INTRODUCTION

Human apolipoprotein C-II (apo C-II) is a 79 amino acid protein secreted primarily by the liver and present in plasma associated with triglyceride-rich lipoproteins and high density lipoproteins (HDL)[1]. As a cofactor for lipoprotein lipase, the enzyme that catalyzes the hydrolysis of triglycerides on lipoproteins, the presence of normal apo C-II is crucial for normal lipoprotein metabolism[2,3].

Familial deficiency of apo C-II is a rare disease inherited as an autosomal recessive trait. Patients homozygous for the disease may present with recurrent abdominal pain, pancreatitis, eruptive xanthomas and lipemia retinalis. In addition they show marked elevation of plasma triglycerides and chylomicrons, decreased LDL and HDL and type I hyperlipoproteinemia phenotype[4,5].

Of the various families with a defect in apo C-II that have been described, an abnormal plasma apo C-II variant has been isolated and the protein defect determined in two kindreds. In apo C-II$_{Toronto}$, the nonfunctional apo C-II protein is present at approximately normal levels and has a different amino acid sequence at positions 69-74 as well as loss of amino acids 75-79 when compared to normal apo C-II[6]. Apo C-II$_{St. Michaels}$ contains a proline instead of a glutamine at position 70 and the abnormal apo C-II sequence is extended 17 residues past the normal carboxyl-terminal amino acid[7]. These abnormalities are most consistent with a base deletion for apo C-II$_{Toronto}$ and a base insertion for apo C-II$_{St. Michael}$ resulting in a subsequent shift of the translation reading frame. The nature of the molecular defect in the other apo C-II deficient families is as yet unknown but it is likely that the syndrome will be found to be a heterogenous disorder at the molecular level.

We have previously described two subjects (brother and sister) affected with the syndrome[8]. Analysis of apo C-II in these patients revealed extremely low levels of apo C-II (0.13 mg/dl and 0.12 mg/dl in the male and female proband, respectively). A variant of apo C-II (apo C-II$_{Padova}$) with lower apparent molecular weight and more acidic isoelectric point was identified in both probands by two-dimensional gel electrophoresis. The marked hypertriglyceridemia and elevation of triglyceride-rich lipoproteins were corrected by the infusion of normal plasma or the injection of a biologically active synthetic 44-79 amino acid residue peptide fragment of apo C-II. Analysis of the apo C-II gene in the patients by restriction enzyme analysis established that the apo C-II gene was present and there were no major insertions or deletions in the apo C-II gene[9].

In the present study we evaluated the molecular defect of the apo C-II$_{Padova}$ family analyzing the presence of apo C-II and the levels of apo C-II mRNA in the hepatocytes in one of the apo C-II deficient probands.

METHODS

Experimental Subject

The apo C-II deficient patient (SF) is 1 of 2 affected individuals from the Padova kindred described previously[8].

RNA Preparation

Liver tissue was obtained from the patient during open abdominal surgery for cholecystectomy, and from control subjects at the time of organ donation. Tissue was stored at -70 °C until used. RNA was isolated utilizing the guanidine thiocyanate method as previously described[10].

Complementary DNA (cDNA) Probes

A 354 base pair AluI restriction fragment of an apo C-II cDNA clone[11] was utilized for our studies. For positive controls, an 850 base pair MspI restriction fragment of an apo A-I clone[12] and a 412 base pair fragment from a DraI and RsaI double digest of a 3'-untranslated ß-actin cDNA clone[13] were used. The ß-actin cDNA clone was kindly provided by Dr. Lawrence Kedes.

Northern and Slot Blot Hybridization Analyses of RNA

Gels for Northern blot analysis were prepared with 1% agarose in the presence of 6% formaldehyde, electrophoresed at 25 volts for 16 hours, and transferred to nitrocellulose filters (Schleicher and Schuell, Inc., Keene, N.H.) as described previously[14]. Eight ug of total RNA were analyzed and gels were stained with ethidium bromide to confirm that equivalent quantities of RNA were loaded in each lane.

For slot blot analysis, serial dilutions of total RNA (3.0, 2.0, and 1.0 μg) were loaded in duplicate onto nylon filters (Gene Screen Plus, NEN/Dupont, Boston, MA) using a slot blot apparatus (Bethesda Research Labs, Gaithersburg, MD.). Baking, prehybridization, and hybridization conditions were as previously described except that hybridization was performed at 42°C for 24 hrs[15]. Filters were auto-radiographed and the blots were quantitated using a laser desitometer (Ultrascan XL, LKB Instruments, Bromma, Sweden). The absorbancy values were normalized with the mean of the values for normal RNA being assigned a value of 1. For re-use, filters were stripped of radiolabelled probe by incubation in 0.1 x SSC (30mM citrate, 0.3M NaCl, pH 7.0)/1% NaDodSO$_4$ at 90°C for one hour.

In Situ RNA Hybridization

Frozen sections were prepared as described for immunohistochemical studies. An AluI cDNA apo C-II fragment was ligated into the SmaI cloning site of PGem 3 vector DNA (Promega, Madison, WI.), and radiolabelled sense and antisense strands were synthesized by using T7 and SP6 DNA polymerases (Promega, Madison, WI.), respectively[16]. In situ RNA hybridization was performed as previously described[16].

Immunohistochemistry

Frozen sections (8μm thick) were prepared from liver biopsies of normal and apo C-II deficient individuals. The liver tissue was embedded in OCT compound (Miles Scientific, Naperville, Il.), sectioned and stored at -70°C until used. Specimens were incubated at room temperature for 30 minutes with apo C-II monoclonal antibodies or an irrelevant monoclonal antibody derived from a neuroblastoma (HSAN 1.2, a gift of C Patrick Reynolds, Naval Medical Research Institute) followed by incubation with the colloidal gold linked secondary antibody (Auro Probe [TM]LM, Jansen Life Sciences Products, Olen, Belgium) at room temperature for 30 minutes. Silver enhancement was as described by the manufacturer (Integrated Separation Systems, Hyde Park, MA). Following counter staining with Hanes' haematoylin, the slides were evaluated for staining with brown or black pigment indicative of a positive reaction.

RESULTS

Northern blot hybridization of total RNA isolated from the liver of 3 normal and our apo C-II deficient patient revealed approximately normal levels of a normal sized apo C-II mRNA (Fig. 1). Stripping and rehybridization of these blots with cDNA probes for apo A-I and β-actin demonstrated equal loading of total liver RNA for all lanes (Fig. 1).

Further quantitation of the apo C-II message level in our patient's liver was performed by slot blot and in situ RNA hybridization analysis.

The data presented in Figure 2 established that the level of hepatic apo C-II mRNA in the patient's liver is approximately normal as determined by slot blot hybridization. Stripping and rehybridization of the slot blot with cDNA probes for apo A-I and ß-actin confirmed that similar quantities of these mRNAs were present in the normal and patient samples. (Data not shown).

Figure 3 shows the _in situ_ RNA hybridization of normal and apo C-II

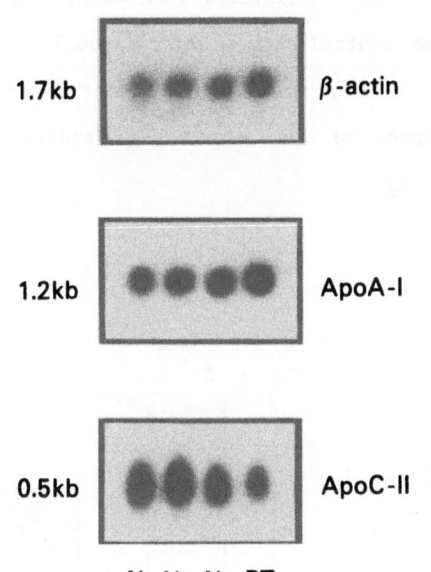

Fig. 1. Northern Blot analysis of RNA from the liver of normal (N1, N2, N3) and patient (PT) hybridized with apo C-II (C), apo A-I (B) and β-actin (A) cDNA probes.

deficient subject liver with an apo C-II antisense RNA probe. In agreement with the results obtained by slot blot hybridization, similar levels of apo C-II message detected by the black grains were identified in both normal (Fig.3B) and patient (Fig.3C) livers. As expected, hybridization of normal liver with apo C-II sense RNA detected no message (Fig.3A).

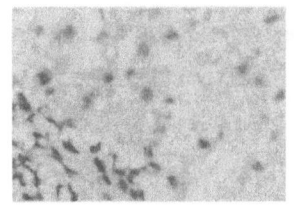 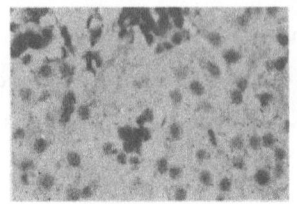 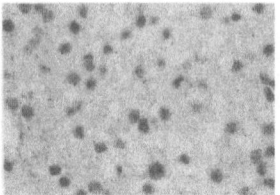

A. Normal Liver   B. Normal Liver   C. Patient Liver

Fig. 2. Quantitation of total hepatic apo C-II mRNA from 2 normal
controls (N1, N2) compared to the apo C-II deficient patient
(PT) by slot blot analysis. All values were normalized to the
mean of the control values being equal to one. Error bars are
standard deviation of the respective mean. Representative
autoradiographs of the slot blot studies are included at the
top of the Figure.

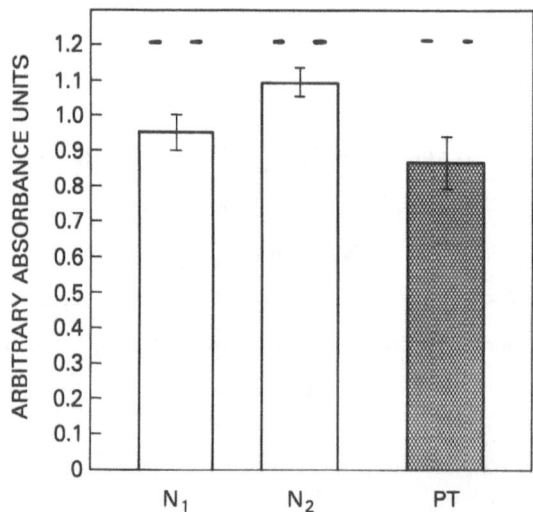

Fig. 3. In situ RNA hybridization of normal (B) and patient (C) livers
hybridized with the apo C-II antisense RNA probe. Normal liver
hybridized with an apo C-II sense RNA probe (A) is included
as a negative control.

IMMUNOHISTOCHEMISTRY FOR
APOLIPOPROTEIN C-II

 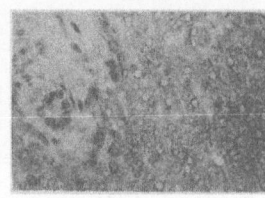 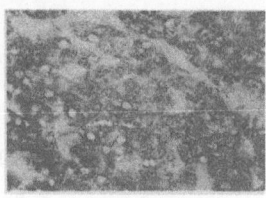

| A. Normal Liver | B. Normal Liver | C. Patient Liver |
| Irrelevant Antibody | Anti apo C-II | Anti apo C-II |

Fig. 4. Immunohistochemical analysis of normal liver incubated with irrelevant antibody (A) and with anti apoC-II monoclonal antibody (B). Normal or slightly increased cytoplasmic staining is present in the liver of the apo C-II deficient patient following incubation with the apo C-II monoclonal antibody (C).

Immunohistochemical analysis of intrahepatic apo C-II was done by using a monoclonal anti C-II antibody and the immunocolloidal gold procedure. Normal liver incubated with an irrelevant antibody showed no staining (Fig.4A). Significant amounts of apo C-II detected by the brown staining within the cytoplasm of hepatocytes, but not within connective tissue, were evident when normal liver was incubated with the monoclonal anti C-II antibody (Fig.4B). Slightly increased amount of cytoplasmic staining is present in the proband's liver (Fig.4C) when compared to normal (Fig.4B). Thus, the amount of C-II apolipoprotein present in the liver of the proband with apo C-II deficiency is not reduced.

DISCUSSION

We have analyzed the DNA, RNA and protein of a patient with a deficiency of apo C-II. Studies indicated that the deficiency of apo C-II in this kindred is most likely due to a small structural defect in the coding portion of the apo C-II gene leading to either defective secretion or enhanced catabolism of the protein.

Thus, previous analysis of the apo C-II DNA by Southern blotting has shown no major rearrangements in the proband's apo C-II gene[9,11]. Northern, slot blot and _in situ_ RNA hybridization studies have confirmed the presence of normal levels of a normal sized apo C-II mRNA in the proband's liver. Immunohistochemical analysis revealed normal to slightly elevated intrahepatic apo C-II content, although plasma apo C-II levels by RIA were markedly reduced to 0.13 mg/dl (normal values = 5.18 ± 0.3 mg/dl). In addition, two-dimensional gel electrophoresis and immunoblotting have previously shown the presence of an abnormal C-II apolipoprotein, apo C-II$_{Padova}$, that exhibits a smaller molecular weight and more acidic pH than normal apo C-II[8]. Thus, this patient has normal quantities of apo C-II mRNA and protein in the liver, but markedly reduced concentrations of an abnormal C-II apolipoprotein in plasma.

These findings are most consistent with a small DNA mutation in
the proband's apo C-II gene that leads to either defective secretion
of the abnormal apo C-II$_{Padova}$ from the hepatocyte or enhanced
catabolism of the abnormal apolipoprotein once secreted into plasma.
With both, abnormal secretion or enhanced catabolism of apo C-II, the
intrahepatic content of apo C-II mRNA and protein is expected, as we
have observed, to be normal, but plasma apo C-II levels would be
decreased. The detection of an electrophoretically abnormal apo C-II
variant, apo C-II$_{Padova}$, indicates that the defect resides within the
coding region of the proband's apo C-II gene.

Further studies are underway to define the precise molecular defect
that leads to abnormal secretion or enhanced catabolism and thus to
deficiency of apo C-II in this kindred.

ACKNOWLEDGEMENTS

This work was partially supported by the National Research Council
of Italy (C.N.R., Progetto Bilaterale 84.0185.04).

REFERENCES

1. A.V. Hospattanakar, T. Fairwell, R. Ronan, H.B. Brewer Jr., Amino
   acid sequences of human apolipoprotein C-II from normal and hyperlipoproteinemia subjects. J. Biol. Chem. 259: 318 (1983).
2. J.C. LaRosa, C.R. Levy, P. Herbert, S.E. Lux, D.S. Fredrickson,
   A specific apoprotein activator for lipoprotein lipase. Biochem.
   Biophys. Res. Commun. 41: 45 (1970).
3. D.M. Bier, R.J. Havel, Activation of lipoprotein lipase by lipoprotein fraction of human serum. J. Lipid Res. 11: 565 (1970).
4. E.A. Nikkila, Familial lipoprotein lipase deficiency and related
   disorders of chylomicron metabolism, in "The Metabolic Basis of
   Inherited Disease," 5th ed. J.B. Stanbury, J.B. Wyngaarden, D.S.
   Fredrickson, J.L. Goldstein, and M.S. Brown editors. McGraw-Hill,
   Inc., New York, 622 (1983).
5. W.C. Breckenridge, J.A. Little, G. Steiner, A. Chow, and M. Poast,
   Hypertriglyceridemia associated with deficiency of apolipoprotein
   C-II. N. Engl. J. Med. 298: 1265 (1978).
6. P.W. Connelly, G.F. Maguire, T. Hofmann and J.A. Little, Structure
   of apolipoprotein C-II$_{Toronto}$, a nonfunctional human apolipoprotein.
   Proc. Natl. Acad. Sci. USA 84: 270 (1987).

7. P.W. Connelly, G.F. Maguire and A.J. Little, Apolipoprotein C-II_{St. Michael}. Familial apolipoprotein C-II deficiency associated with premature vascular disease. J. Clin. Invest. 80: 1597 (1987).

8. G. Baggio, E. Manzato, C. Gabelli, R. Fellin, S. Martini, G. Baldo Enzi, F. Verlato, M.R. Baiocchi, D.L. Sprecher, M.L. Kashyap, H.B. Brewer, Jr. and G. Crepaldi, Apolipoprotein C-II Deficiency Syndrome. J. Clin. Invest. 77: 520 (1986).

9. S.E. Humphries, L. Williams, O. Myklebost, A.F. Stalenhoef, P.N.M. Demacker, G. Baggio, G. Crepaldi, D.J. Galton and R. Williamson, Familial Apolipoprotein C-II Deficiency: A Preliminary Analysis of the Gene Defect in Two Independent Families. Hum. Genet. 67: 151 (1984).

10. J.A. Chirgwin, A.E. Przybyla, R.J. MacDonald and W.J. Rutter, Isolation of Biologically Active Ribonucleic Acid From Sources Enriched in Ribonuclease. Biochemistry 18: 5294 (1979).

11. S.S. Fojo, S.W. Law, D.L. Sprecher, G. Baggio and H.B. Brewer, Jr., Analysis of the apoC-II Gene in apoC-II Deficient Patients. Biochem. Biophys. Res. Commun. 124: 308 (1984).

12. S.W. Law and H.B. Brewer, Jr., Nucleotide Sequence and the Encoded Amino acids of Human Apolipoprotein A-I mRNA. Proc. Natl. Acad. Sci. USA 81: 66 (1984).

13. P. Ponte, S.-Y. Ng, J. Engel, P. Gunning and L. Kedes, Evolutionary Conservation in the Untranslated Regions of Actin mRNAs: DNA Sequence of a Human Beta-Actin. Nucleic Acid. Res. 12: 1687 (1984).

14. J. Meinkoth and G. Wahl, Hybridization of Nuclear Acids Immobilized on Solid Supports. Anal. Biochem. 138: 267 (1984).

15. R.S. Ross, R.E. Gregg, S.L. Law, J.C. Monge, S.M. Grant, K. Higuchi, Y.J. Triche, J. Jefferson and H.B. Brewer, Jr., Homozygous Hypobetalipoproteinemia: A Disease Distinct From Abetalipoproteinemia at the Molecular Level. J. Clin. Invest. 81: 590 (1988).

16. M.E. Harper, L.M. Marselle, R.C. Gallo and F. Wong-Staal, Detection of Lymphocytes Expressing Human T-Lymphotropic Virus Type III in Lymph Nodes and Peripheral Blood From Infected Individuals by in Situ Hybridization. Proc. Natl. Acad. Sci. USA 83: 772 (1986).

BIOCHEMICAL ASPECTS AND MOLECULAR STUDY OF A CASE OF APO CII DEFICIT

°Pepe G.,°Crecchio C.,°Tullo A., *Mogavero A.M.,*De Tommaso M.
and *Capurso A.

°Centro SMME-CNR, Dipartimento Biochimica Biologia Molecolare
*Cattedra Gerontologia e Geriatria, Istituto Medicina Clinica
Università di Bari, Italy

INTRODUCTION

The apolipoprotein CII (Apo CII), as major component of VLDL and
HDL and physiological activator of the Lipoprotein Lipase (LPL) enzyme, is
largely involved in the metabolism of lipids. Thus variant forms of this
protein or its plasmatic deficit are responsible for unbalanced lipid catabolism with clinical consequences: patients with familiar Apo CII deficiency
develop severe hypertrigliceridaemia.

The mature protein of 79 aa with known sequence[1] is codified by a
gene localized on chromosome 19[2] and it is secreted from epatic and intestinal cells into the plasma. The organization of the codifying genomic region
has been characterized in normal human cells[3].

In the present report we describe the case of two patients totally
lacking of plasmatic Apo CII, although the protein is immunodetectable
within their intestinal cells. Since biochemical evidences suggest that the
synthesis of Apo CII occours in almost normal amount, at least in this
tissue, we have investigated the origin of the defect at molecular level,
through a detailed analysis of the genic organization of one of the affected
subjects.

MATERIALS AND METHODS

Lipoprotein classes were separated by sequential ultracentrifugation
according to Havel et al.[4]. Apoprotein determination was performed by
radioimmunodiffusion using a commercial polyclonal monospecific antibodies
(Daiiki, Tokyo). Bidimensional electrophoresis and immunoblot analysis were
carried out according to Burnette[5] and Neville[6] respectively. Analytical
isoelectric focusing was performed on 7.5% polyacrylamide gel according to
Catapano et al.[7] with 2% ampholines, pH 4-6. The immunofluorescence on
intestinal mucosa cells was performed with fluorescin-labelled rabbit anti
human Apo CII serum, according to Cherry et al.[8].

For the hybridization experiments the DNAs were prepared from peripheral blood by phenol-clorophorm extraction, completely digested with
5U/µg of different enzymes (Eco RI, Bam HI, Pst I, Taq I) and blotted

Table 1. Plasmatic cholesterol, tryglicerides and
apoprotein content (mg/dl).

	CH	TG	AI	AII	B	CII	CIII	E
P.I.	137	806	95	26	69	---	12.3	6.2
P.G.	140	604	92	22	67	---	12.0	7.5
Father	207	224	125	33	86	1.3	12.7	5.8
Mother	192	124	130	39	84	3.5	8.9	4.8

according to Southern[9]. The probe used in all experiments derived from a
full length Apo CII cDNA of 500pb (gift of Dr. Sidoli) nick-translated in
standard conditions. The genomic library was constructed in EMBL3 lambda
vector using for packaging the Stratagene Gegapack plus. From 5×10^5
recombinant placques ten positive clones were selected after secondary
and terziary screening with the same Apo CII cDNA probe. They were purified,
DNA was prepared with alkaline procedure[10] and the inserts were excised,
digested with appropriate restriction enzymes and subcloned in pUC8 to be
sequenced.

RESULTS AND DISCUSSION

The two patients P.I. and P.G. and their relatives were tested for
the lipid plasmatic content. The clinical data showed very high tryglicerides values and complete absence of Apo CII in the probands, whereas both
parents and some relatives had half value of this apoprotein (Table 1).
According to these data the conditions of homozygosis and heterozygosis
were ascribed to the siblings and their parents respectively.

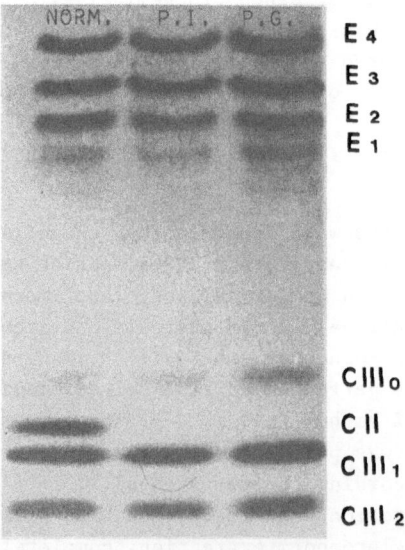

Fig. 1. Analytical isoelectrofocusing of delipidated Apo-VLDL.

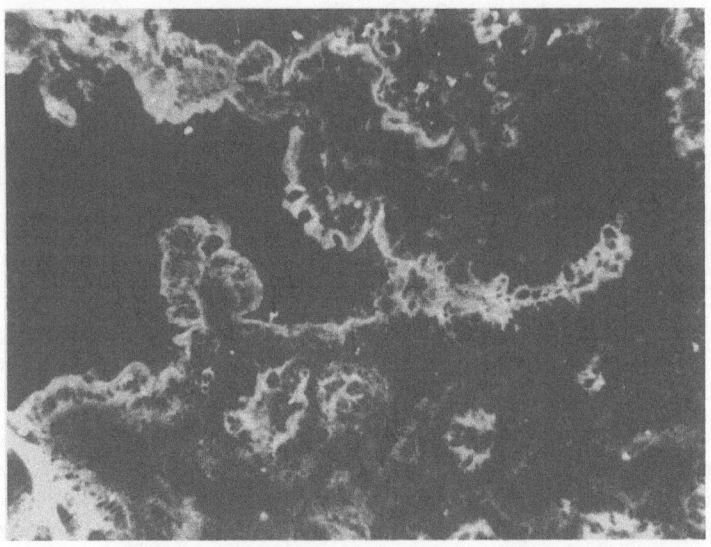

Fig. 2. Immunofluorescence of P.G. intestinal mucosa.
250X.

The total plasmatic deficiency of Apo CII of both probands was also
demonstrated by isoelectric focusing, bidimensional electrophoresis and
immunoblot analysis, as described in Materials and Methods. Fig.1 shows an
isoelectrofocusing slab gel of apo-VLDL of the two probands compared with
a normal pattern.

On the basis of these results we must conclude that no Apo CII is
detectable in the plasma of P.I. and P.G. However immunoreactive Apo CII
has been found in the intestinal cells of the patients with amount equal
or even higher than the normal control (Fig.2).

These data together suggest that we are studying a variant form of
protein synthesized at more or less normal level. However a genic alteration
affects in some way the secretion and/or the stability within the synthe-
sizing cells or in the plasma. This can prevent the assembling with lipidic
components of lipoproteins and their normal utilization.

In order to clarify the type and the localization of the mutation
responsible for the defect we have investigated the genic organization of
the patients, focussing our attention on one of the two siblings, the girl
P.I.

The DNA of the proband, parents and normal subject as control,
restricted by Eco RI,were hybridized using the Apo CII cDNA probe. As shown
in Fig.3 only one hybridization band appears on the autoradiography,
corresponding to the Rco RI fragment of 3500bp, which, according to Wei
et al.[3], should contain the entire normal gene (4 exons and 3 introns) for
the Apo CII protein. The position of the radioactive band is in favour of
a normal genic organization in the patient and support the idea that no
 gross alteration took place in it. This is also confirmed by the fact that
up to now we did not find a RFLP related with the defect with any of the

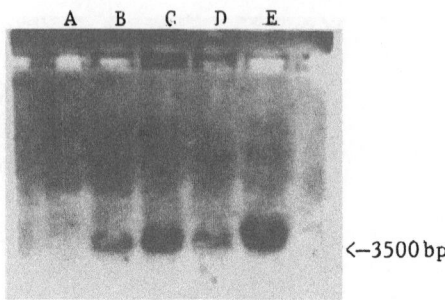

A B C D E

<--3500 bp

Fig. 3. Hybridization with Eco restricted DNAs: A= xHindIII;
B=normal; C=father; D=mother; E=P.I.(different amount).

enzymes tested. Digestion by Taq I and hybridization with the same probe
shows the unique band of 3800bp, which, according to Humphries[11], is commun
to the 60% of the normal caucasian population (Fig.4, lanes E,F). Moreover
the double digested Eco RI-Pst I DNA from P.I. reveals bands corresponding
in size to that expected from the map of the normal gene[3](Fig.4, lanes C,D).

Thus it seems possible to exclude in our patient a large rearrangement.
The defect should depend on a small gene mutation which could partly affect·
the correct splicing or introduce a premature stop codon(as reported by
Foyo in its cases of Apo CII deficiency (this issue)),or aminoacid changes.

In order to investigate this point we have constructed the genomic
library of P.I. The positive clones have been further characterized after
excision of the inserts and secondary digestion. Sequencing in progress of
these fragments subcloned in plasmid will reveal the origin of the defect.

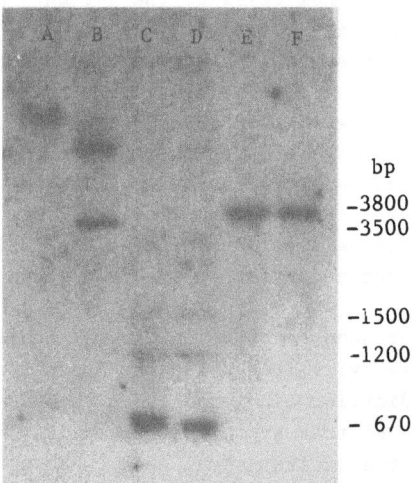

A B C D E F

bp
-3800
-3500

-1500
-1200

- 670

Fig. 4. Hybridization with differently digested DNAs: A= xEcoxHind;
B=P.I.xEco; C=normalxEcoxPst; D=P.I.xEcoxPst; E=normalxTaq;
F=P.I.xTaq.

ACKNOWLEDGEMENTS

Research supported by MPI grant and by Regione Puglia grant: Ricerca
Sanotaria Finalizzata

REFERENCES

1) Jackson RL., Baker HN., Gilliam EB., Gotto A. Jr. Primary structure of
very low density apolipoprotein CII of human plasma. Proc. Natl. Acad.
Sci. USA 74: 1942 (1977).

2) Brook JD., Shaw DJ., Meredith L., Bruns GAP., Harper PS. Localization
of genetic markers and orientation of the linkage group on chromosome 19.
Human Genet. 68: 282 (1984).

3) Wei C-F., Tsao Y-K., Robberson DL., Gotto A. Jr., Brown K., Chan L.
The structure of the human apolipoprotein C-II gene. J. Biol. Chem.
28: 15211 (1985).

4) Havel RJ., Eder HA., Bragdon JH. The distribution and chemical composition of ultracentrifugally separated lipoproteins in human serum.
J. Clin. Invest. 34: 1345 (1955).

5) Burnette WN. Western blotting: electrophoretic transfer of proteins from
sodium dodecyl sulfat-polyacrilamide gels to unmodified nitrocellulose
and radiographic detection with antibody and radioiodinated protein A.
Biochem. 112: 195 (1981).

6) Neville DM. Molecular weight determination of protein-dodecylsulfate
complexes by gel electrophoresis in a discontinuous buffer system.
J. Biol. Chem. 246: 6328 (1971).

7) Catapano AL., Jackson RL., Gilliam EB., Gotto A.M. Jr., Smith LC.
Quantification of Apo CII and Apo CIII of human very low density lipoproteins by analitical isoelectrofocusing. J. Lipid Res. 19: 1047 (1978).

8) Cherry WB., Goldman M., Carski TR. Fluorescent antibody techniques in
the diagnosis of communicable diseases. U.S. Public Health Service Publ.
729 (1960).

9) Southern EM. Detection of specific sequences among DNA fragments separated by gel electrophoresis. J. Mol. Biol. 98: 503 (1975).

10) Birnboim HC., Doly J. A rapid extraction procedure for screening
recombinant plasmid DNA. Nucl. Acids Res. 7: 1513 (1979).

11) Humphries SE., Williams L., Myklebost O., Stalenhoef AFH., Demacker PNM.,
Baggio G., Crepaldi G., Galton DJ. Williamson R. Familial apolipoprotein
CII deficiency: a preliminary analysis of the gene defect in two
independent families. Human Genet. 67: 151 (1984).

THE MOLECULAR BASIS OF THE DEFECT IN FAMILIAL COMBINED
APOLIPOPROTEINS AI AND CIII DEFICIENCY

Sotirios K. Karathanasis

Laboratory of Molecular and Cellular Cardiology
Children's Hospital, and Department of Pediatrics
Harvard Medical School, Boston, MA 02115

INTRODUCTION

A key feature in atherosclerosis is the progressive accumulation of cholesterol in cells of the arterial wall. Animal cells control their cholesterol content through the integration of the pathways involved in cellular biosynthesis, uptake and secretion of cholesterol. Secreted cholesterol is transported through the plasma to the liver via a series of reactions collectively termed reverse cholesterol transport. A key step in reverse cholesterol is the efflux of cholesterol from cell membranes to a species of high density lipoproteins (HDL). The major protein constituent of HDL is apolipoprotein AI (apoAI). ApoAI activates lecithin:cholesterol acetyltransferse (LCAT), an enzyme involved in esterification of cholesterol on HDL particles. This esterification reaction is thought to be essential for net transport of cholesterol from cells into HDL (reviewed in refs. 1 and 2). These considerations, taken together, imply that deficiency or chronic reduction in plasma HDL and apoAI levels may result in increased risk for atherosclerosis. In support of this possibility, a large number of epidemiological studies have revealed a strong inverse correlation between plasma HDL and apoAI levels and atherosclerosis (reviewed in ref. 3). However, until recently, direct evidence for a cause and effect relationship between plasma HDL/apoAI levels and risk for atherosclerosis was not available. An approach to establish such a relationship is to show that at least in some cases, the only risk factor associated with atherosclerosis is primary (i.e. genetic) HDL deficiency.

Multiple genetic factors (i.e. genes) influence plasma HDL levels. Among them the apoAI gene is of particular interest because the protein product of this gene (i.e. apoAI) is involved in the control of both HDL structure by its participation as the major protein constituent in HDL particles, and HDL metabolism by its involvement in LCAT activation. Therefore, inactivation of the apoAI gene may result in drastic reduction in plasma HDL levels, severe dysfunction of reverse cholesterol transport, excessive accumulation of cholesterol in peripheral tissues and premature atherosclerosis. Certain rare patients present clinical manifestations similar to those expected when the apoAI gene is inactivated. For example, in two sisters with extraordinary low plasma HDL levels, deficiency in both plasma apoAI and another apolipoprotein, apolipoprotein CIII (apoCIII), skin and tendon xanthomas, corneal clouding, and severe premature atherosclerosis at a very early age (i.e. late twenties) have been previously described (4). Because neither apoAI nor apoCIII are detectable in the plasma of these patients and both of these proteins are present approximately at half the normal levels in the plasma of the first degree relatives of these patients, this disease entity has been termed familial combined apoAI and apoCIII deficiency (FC AI/CIII deficiency).

In previous studies we have shown that the genes coding for apoAI, apoCIII and another apolipoprotein, apolipoprotein AIV (apoAIV) are closely linked and tandemly organized within a 15 kilobase (kb) DNA segment in the long arm of human chromosome 11. The apoAI and apoCIII genes are transcribed in opposite directions and the DNA region between them is approximately 2.6 kb in length (see Fig. 1, Chapter , page , in this book). Cloning and characterization of this DNA segment provided DNA probes suitable for structural characterization of all three of these genes in the genome of any individual. Because it seemed possible that the defect in FC AI/CIII deficiency is due to inactivation of the apoAI gene, we used these DNA probes to study the structural organization of all three of these genes in the genome of these patients and their first degree relatives. In this chapter I will summarize the concepts, experimental approaches and some of the results we employed to document that the molecular defect in FC AI/CIII deficiency is due to a DNA inversion which results in aberrant expression of both apoAI and apoCIII genes in these patients. These findings establish a cause and effect relationship between plasma HDL and apoAI levels and atherosclerosis, and provide strong presumptive evidence for the concept that HDL has anti-atherogenic potential.

1. Rearrangement of the apoAI gene in the genome of patients with familial combined apoAI and apoCIII deficiency

The structure of the apoAI gene in patients with FC AI/CIII deficiency was initially studied by genomic blotting analysis. Specifically, chromosomal DNA prepared from peripheral blood lymphocytes of these patients was digested with various combinations of restriction enzymes, electrophoresed in agarose gels, transferred onto nitrocellulose filters and the filters were hybridized with an apoAI cDNA probe. This probe contains sequences corresponding to the fourth exon of the human apoAI gene (ref. 5, see also Fig. 2, probe II). Chromosomal DNA from

Table I. Sizes (in kb) of genomic restriction fragments hybridizing to an apoAI cDNA probe (see text).

Restriction Enzyme	Normal	FC AI/CIII Deficient Patient
EcoRI	13	6.5
BamHI	12	11
HindIII	15	15
PstI	2.2	0.95
EcoRI+BamHI	12	6.5
EcoRI+HindIII	5.5	6.5
BamHI+HindIII	5.5	11
PstI+HindIII	1.7	0.95
EcoRI+BamHI+HindIII	5.5	6.5

normal individuals was also similarly analyzed. For each enzymatic
digestion the sizes of DNA fragments corresponding to hybridization bands
in the resulting autoradiograms were determined (see Fig. 1 in ref. 5).
The results have been compiled in Table I. They show that, with DNA from
normal individuals, every enzymatic digestion results in DNA fragments
with sizes compatible to those expected based on the restriction map of
the apoAI gene and its flanking DNA sequences (see Fig. 1, Chapter ,
page , this book). In marked contrast, these results show that with DNA
from the FC AI/CIII deficient patients most of the enzymatic digestions
result in DNA fragments with sizes different than those expected.

These results indicated that the structure of the apoAI gene in the
genome of these patients is different from that of normal individuals.
This difference cannot be explained by a single restriction site
polymorphism because multiple restriction sites seem to be affected. In
addition, since the apoAI cDNA probe hybridizes to the patients' DNA it
can be concluded that this difference cannot be explained by deletion of
the apoAI gene sequences corresponding to this probe. It therefore
appears that the apoAI gene sequences in the genome of these patients
have been rearranged.

2. The apoAI gene rearrangement cosegregates with reduced apoAI, apoCIII and HDL levels in the plasma of family members of patients with familial combined apoAI and apoCIII deficiency

The first degree relatives of the patients with FC AI/CIII
deficiency, although asymptomatic for atherosclerosis, have approximately
half normal apoAI, apoCIII and HDL levels (4,5). It was therefore of
interest to determine whether these individuals are heterozygotes for the
apoAI gene rearrangement. To study this possibility, chromosomal DNA
from the father, mother, brother and offspring of these patients was
subjected to genomic blotting analysis by digestion with the enzyme EcoRI
and hybridization with the apoAI cDNA probe described in the previous
section. The restriction enzyme EcoRI was chosen for this analysis

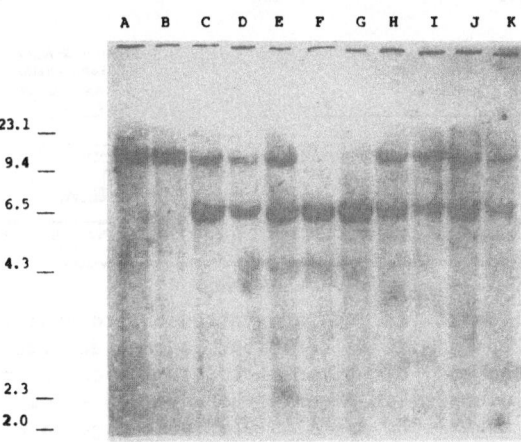

Fig. 1.   Genomic blotting analysis of EcoRI-digested DNA from a normal
          individual (A), the FC AI/CIII deficient probands(F) and (G),
          their maternal grandfather (B), father (C), mother (D), brother
          (E), and the son (H) and daughter (I) of proband (G), and the
          son (J) and daughter (K) of proband (F) using an apoAI cDNA
          probe (see text). The resulting autoradiogram is shown. DNA
          molecular weight standards (in kb) are shown by numbers on the
          left side of the autoradiogram.

because it permits a clear distinction between the sizes of DNA fragments
in hybridization bands obtained with the normal (13 kb) and the
rearranged (6.5 kb) apoAI genes (see Table I). The resulting
autoradiogram is shown in Fig. 1. It shows that every first degree
family member of these patients is a heterozygote for the normal and
rearranged apoAI genes.

These results indicated that the apoAI gene rearrangement
cosegregates with reduced plasma apoAI, apoCIII and HDL levels and
suggested that this gene rearrangement may have inactivated both apoAI
and apoCIII genes in these patients.

3. The apoAI gene rearrangement is due to inversion of a DNA segment containing portions of the apoAI and apoCIII genes including the DNA region between them

To further characterize the apoAI gene rearrangement we constructed
a genomic library using peripheral blood DNA from one of the FC AI/CIII
deficient patients (6). Screening of this library with an apoAI gene
probe resulted in isolation of several positive clones. One of them was
further purified and its recombinant DNA insert was mapped with various
restriction enzymes. The organization of apoAI and apoCIII gene
sequences in this clone was studied as follows: DNA prepared from this
clone was digested with several combinations of restriction enzymes,
electrophoresed in agarose gels, blotted onto nitrocellulose filters and
the filters were hybridized with probes prepared using DNA fragments
derived from different regions of the normal apoAI and apoCIII genes.
Specifically, probe I was prepared using a 0.45 kb PstI-HindIII DNA
fragment spanning the "TATA box", exon 1, intron 1, exon 2 and part of

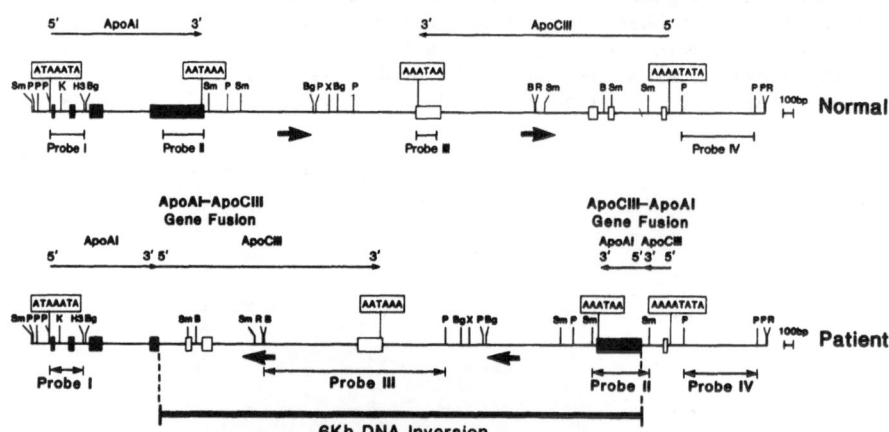

Fig. 2.   Restriction enzyme maps and intron-exon organization of the
          normal and FC AI/CIII deficient patient derived apoAI and
          apoCIII genes. Direction of transcription (arrows), TATA boxes
          and polyadenylation signals (boxed), and exons in the apoAI
          (filled boxes) and apoCIII (open boxes) genes are indicated.
          The scale of the maps are shown by 100-bp bars. Restriction
          sites are: X, Xba I; H3, HindIII; Bg, BglII; K, Kpn I; P, Pst I;
          B, BamHI; Sm, Sma I; and R, EcoRI. DNA fragments used for
          probes are indicated by lines below the map of the normal apoAI
          and apoCIII genes. Restriction fragments with hybridization
          homology to these probes (double headed arrows) and the 6.0 kb
          DNA inversion (thick line) are indicated below the map of the
          patient derived genes. Alu I repetitive DNA elements are showhn
          by thick arrows.

intron 2 in the apoAI gene; probe II was prepared using a 0.5 kb apoAI
cDNA fragment spanning a portion of exon 4 in the apoAI gene; Probe III
was prepared using a 0.25 kb apoCIII cDNA fragment spanning a portion of
exon 4 in the apoCIII gene; and probe IV was prepared using a 0.9 kb
PstI-PstI DNA fragment located immediately upstream to the 5' end of the
"TATA box" of the apoCIII gene (see Fig. 2, normal). The resulting
autoradiogram (see Fig. 1B in ref. 6) showed that the obtained
hybridization patterns are identical to those obtained by genomic
blotting analysis of blood DNA from one of these patients using the same
probes (see Fig. 1 in ref. 5). These results indicated that the
arrangement of sequences within the recombinant DNA insert in this clone
represents the authentic arrangement of these sequences within the
genome of these patients. In addition, these results facilitated mapping
of restriction fragments with hybridization homology to each of these
probes within the restriction map of the recombinant DNA insert in this
clone (Fig. 2, patient). Comparison of the relative location of these
probes between the normal genome and the patient derived clone revealed
striking differences. Specifically, the arrangement of these probes in
the normal genome is I-II-III-IV while in the clone it is I-III-II-IV
(Fig. 2). Furthermore, the restriction maps to the left of probe I and
to the right of probe IV are identical between the normal genome and the
patient derived clone, while the restriction maps of the DNA region
between these probes are identical between the normal genome and the
patient derived clone only if one of these two DNA regions is inverted
(Fig. 2).

   These results indicated that an approximately 6.0 kb DNA segment
containing portions of the 3' sequences in the apoAI and apoCIII genes
including the DNA region between these genes is inverted in the genome of
the FC AI/CIII deficient patients. In addition, these results suggested
that the remaining 5' sequences, promoters and 5' flanking regions of
these genes are not affected by this DNA inversion.

4. The DNA inversion is confined within the apoAI-apoCIII gene region

   The determine whether the DNA inversion in the genome of the FC
AI/CIII deficient patients has altered the structure of sequences

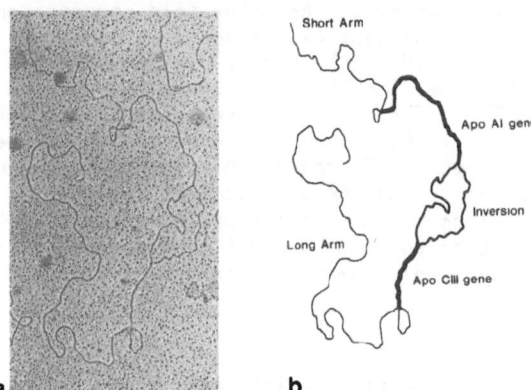

a   b

Fig. 3.   (A) Electron micrograph of the heteroduplex formed between
          genomic clones derived from the genome of a normal individual
          and a patient with FC AI/CIII deficiency.   (B) Drawing of the
          heteroduplex in A showing similarities (thick lines) and
          differences (thin lines) of the apoAI-apoCIII gene region of the
          recombinant DNA inserts in these clones.   The junctions between
          the arms of cloning vector and the DNA inserts in these clones
          are the indicated loops.

flanking the apoAI and apoCIII genes, first we studied the structure of the apoAIV gene in the genome of these patients. Genomic blotting restriction mapping analysis using an apoAIV gene probe showed that the hybridization patterns obtained with DNA from these patients are identical to those obtained with DNA from normal individuals (see Fig. 5 in ref. 7). In parallel experiments we isolated a genomic clone containing the apoAI and apoCIII genes in their normal arrangement (Fig. 2, normal, see also ref. 6). This clone also contains 8.5 kb and 2.5 kb of the 5' flanking regions of the apoAI and apoCIII genes, respectively. The patient derived genomic clone (see previous section) contains, in addition to the DNA inversion, 6.5 kb and 5.0 kb of the 5' flanking regions of the apoAI and apoCIII genes, respectively. The recombinant DNA inserts in both of these clones have the same orientation with respect to the arms of the cloning vector. The structure of these DNA inserts was compared to each other by using the corresponding clones for heteroduplex mapping analysis. The resulting electron micrograph is shown in Fig. 3 (see also ref. 6). It shows that, with the exception of the 6 kb DNA inversion, the remaining sequences are very similar between these clones.

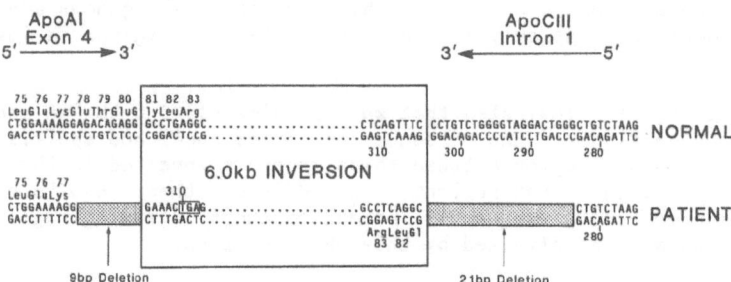

Fig. 4. Alignment of nucleotide sequences at the junctions of the DNA inversion in the FC AI/CIII deficient patient derived clone (lower sequence) with the corresponding sequences in the normal human genome (upper sequence). Direction of transcription is indicated by arrows. The amino acid sequence encoded by the portion of apoAI gene exon 4 shown is numbered according to the first amino acid in the mature plasma apoAI. A termination codon in frame with the apoAI translation frame is indicated by a boxed TGA. The nucleotide sequence in the portion of apoCIII gene intron 1 shown is numbered according to the first nucleotide at the 5' end of this intron in the normal human apoCIII gene. The DNA segment that is inverted in the patient derived clone and the corresponding segment in the normal human genome are boxed. The nucleotide sequences deleted in the patient derived clone are indicated by stippled boxes.

Taken together these results indicated that the structure of at least 6.5 kb of sequences flanking the 5' end of the apoAI gene and 16.5 kb of sequences flanking the 5' end of the apoCIII gene, which include the entire apoAIV gene, is very similar between the FC AI/CIII deficient patients and normal individuals. We therefore concluded that the DNA inversion in the genome of these patient is confined within the apoAI-apoCIII gene region and that these patients have a structurally and possibly functionally normal apoAIV gene.

5. The DNA inversion results in reciprocal fusion of the apoAI and apoCIII gene transcriptional units

Since the DNA inversion in the genome of the FC AI/CIII deficient patients seems to have altered the structural organization of DNA sequences within the apoAI and apoCIII genes, it was of interest to determine whether this alteration could have resulted in inactivation of both these genes and thus could account for the combined deficiency of apoAI and apoCIII in the plasma of these patients. For this reason, the sequences at the junctions of the DNA inversion in the patient derived genomic clone were determined (6). The results showed that as a consequence of this DNA inversion, the 3' end of a portion of the apoAI gene exon 4 is linked to the 5' end of a portion of the apoCIII gene intron 1, while the 3' end of a portion of the apoCIII gene intron 1 is linked to the 5' end of a portion of the apoAI gene exon 4 (Fig. 2). The sequences flanking the junctions of this DNA inversion were aligned and compared with the corresponding sequences in the normal human genome (Fig. 4). This comparison showed that 9 nucleotides in the apoAI gene exon 4 (AGACAGAGG) and 21 nucleotides in the apoCIII gene intron 1 (CCCAGTCCTACCCCAGACAGG) have been deleted in the patient derived clone (Fig. 4). Therefore, this DNA inversion results in fusion of nucleotide 1 in codon 78 of the message translated into the mature apoAI amino acid sequence with nucleotide 305 in the apoCIII gene intron 1 (apoAI-apoCIII fusion) and fusion of nucleotide 283 in the apoCIII gene intron 1 with nucleotide 2 in codon 81 of the message translated into the mature apoAI amino acid sequence (apoCIII-apoAI fusion) (Fig. 4). It also should be noted that translation of the apoAI-apoCIII fusion, according to the frame of apoAI, results in a termination codon located two codons 3' to codon 77 in apoAI (boxed TGA in Fig. 4).

These results indicated that because of the close physical linkage and convergent transcription of the apoAI and apoCIII genes in the normal genome, the DNA inversion in the genome of the FC AI/CIII deficient patients has resulted in reciprocal fusion of the apoAI and apoCIII gene transcriptional units. In addition, these results suggested that as a consequence of this gene fusion, the expression of both of these genes may be altered in these patients.

6. The apoAI-apoCIII and apoCIII-apoAI gene fusions are transcribed into stable mRNAs with sequences representing fusions of the normal apoAI and apoCIII mRNAs

To determine whether the apoAI-apoCIII and apoCIII-apoAI gene fusions can be expressed into stable mRNAs and whether these mRNAs could be translated into proteins, DNA fragments containing these gene fusions were isolated from the patient derived genomic clone (see section 3 above) and subcloned into a plasmid vector (pUC9-SV) containing a powerful transcriptional enhancer derived from the simian virus 40 (SV40). Similarly, DNA fragments containing the normal apoAI and apoCIII genes were isolated from a genomic clone that contains these genes in their normal arrangement and were also subcloned in pUC9-SV. The resulting plasmid gene constructs were transfected into cultured monkey kidney cells (COS-1 cells) and total RNA prepared from these cells was studied by electrophoresis in agarose-formaldehyde gels, blotting onto nitrocellulose filters and hybridization with the apoAI cDNA probe II and apoCIII cDNA probe III (see Fig. 2). The resulting autoradiograms showed that expression of the constructs containing the normal apoAI and apoCIII genes results in mRNAs with sizes identical to the sizes of adult human liver apoAI and apoCIII mRNAs respectively (see Fig. 4 in ref. 6). In sharp contrast, expression of the apoAI-apoCIII gene construct results in a mRNA with homology to probe III but not to probe II and a size that is

400-500 nucleotides larger than adult human liver apoCIII mRNA. The
expression of the apoCIII-apoAI gene fusion, on the other hand, results
in a mRNA with homology to probe II but not to probe III and a size that
is 400-500 nucleotides shorter than adult human liver apoAI mRNA (see
Fig. 4 in ref. 6)

The apoAI-apoCIII gene fusion contains the "TATA box", exon 1,
intron 1, exon 2, intron 2, exon 3, intron 3, and part of exon 4 in the
apoAI gene, and part of intron 1, exon 2, intron 2, exon 3, intron 3 and
all of the exon 4 in the apoCIII gene (Figs. 2 and 5). The intron-exon
organization of this gene fusion and the hybridization of its transcript
to the apoCIII but not apoAI cDNA probes indicates that its transcription
begins at the apoAI gene exon 1 and extends through the apoCIII gene exon
4. Therefore, the nucleotide sequence of this transcript represents a
fusion of the nucleotide sequences present in the normal apoAI and
apoCIII mRNAs. Similarly, the apoCIII-apoAI gene fusion contains the
"TATA box", exon 1 and part of intron 1 in the apoCIII gene and a portion
of exon 4 in the apoAI gene (Figs. 2 and 5). The intron-exon
organization of this gene fusion and the hybridization of its transcript
to the apoAI but apoCIII cDNA probes indicates that its transcription
begins at the apoCIII gene exon 1 and extends through the apoAI gene exon
4. Thus the nucleotide sequence of this gene transcript represents a
fusion of the nucleotide sequences present in the normal apoCIII and
apoAI mRNAs.   Subsequent characterization of the structure of these
transcripts by S1 nuclease and primer extension analysis confirmed these
findings and indicated that the apoAI-apoCIII gene fusion mRNA contains
the apoAI gene exons 1, 2 and 3 and part of exon 4 as well as the apoCIII
gene exons 2, 3 and 4. In addition, this mRNA contains the entire
portion of apoCIII gene intron 1 present in this gene fusion. The
apoCIII-apoAI gene fusion mRNA, on the other hand, contains the apoCIII
gene exon 1 and the 3' portion of the apoAI gene exon 4 (Fig. 5, see also
Fig. 5 in ref. 6).   Since the codons for initiation of translation of
the apoAI and apoCIII mRNAs are located within exon 2 in the
corresponding genes and the apoAI-apoCIII gene fusion mRNA contains both
these exons, the initiation codons for apoAI (upstream initiation codon)
and apoCIII (downstream initiation codon) are both present in the apoAI-
apoCIII fusion mRNA (Fig. 5). Translation of this mRNA starting at its
upstream initiation codon runs into a termination codon located two
codons 3' to codon 77 in apoAI (see Fig. 4). However, translation of
this mRNA starting at its downstream initiation codon results in the
intact normal apoCIII amino acid sequence. Since the apoAI and apoCIII

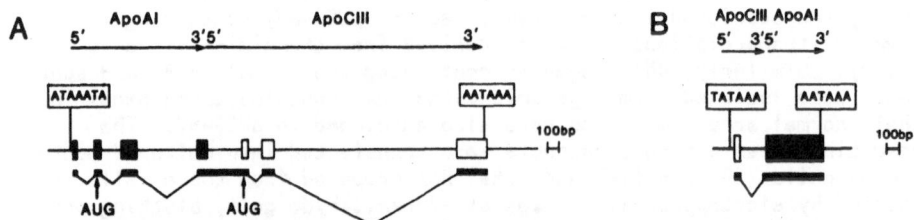

Fig. 5. Splicing patterns of the apoAI-apoCIII (A) and apoCIII-apoAI (B)
         gene fusion transcripts. A map of the organization of these
         genes is indicated as in Fig. 2. The sequences present in the
         corresponding mRNAs (thick lines) and the natural translation
         initiation codons (AUG) for apoAI and apoCIII are shown below
         the map.

deficient patients lack apoCIII in their plasma, it is possible that apoCIII is being synthesized in vivo, but disturbance(s) of the lipoprotein metabolism due to the absence of apoAI results in hypercatabolism of this protein. It is more likely, however, that initiation of translation of this mRNA occurs with very low frequency in its internal (downstream) initiation codon. On the other hand, the apoCIII-apoAI gene fusion mRNA does not contain any of the natural initiation codons for apoAI and apoCIII (Fig. 5). However, the first AUG proximal to the 5' end of this mRNA coincides with codon 86 in apoAI mRNA (8). This AUG is located within a nucleotide sequence context (AGGAGAUGA) that may allow its utilization by ribosomes as an initiation codon. Thus, translation of this mRNA would result in a protein containing the 86th to 243rd amino acid portion of apoAI. Therefore, this protein would contain the amphipathic amino acid repeats in apoAI (see chapter , page , this book) but not its signal peptide and may be accumulated and/or degraded intracellularly.

These results indicated that the apoAI-apoCIII and apoCIII-apoAI gene fusions in the genome of the FC AI/CIII deficient patients can be transcribed into stable mRNAs with nucleotide sequences representing fusions of the sequences present in the normal apoAI and apoCIII mRNAs. In addition, these results indicated that neither of these mRNA fusions can be translated into an intact normal apoAI amino acid sequence and that although one of these mRNA fusions, that is the apoAI-apoCIII mRNA fusion, can be translated into an intact normal apoCIII amino acid sequence, this would require that translation is initiated at an internal initiation codon which may be a very low frequency event. We therefore concluded that the DNA inversion in the genome of these patients has resulted in aberrant expression of both apoAI and apoCIII.

CONCLUSIONS AND FUTURE PERSPECTIVES

Reciprocal recombination events leading to DNA inversions are essential for switching mechanisms in prokaryotes (9,10) and for assembly of functional immunoglobulin (11) and T-cell receptor genes in eukaryotes (12). DNA segments targeted for inversion are usually flanked by characteristic DNA repeats which may serve as "joining signals". The presence of such "joining signals" at the junctions of two different t(11;14) chromosomal translocations of B-cell neoplasms (13), a mouse nonproductive immunoglobulin kappa gene rearrangment (14) and a DNA inversion within the β-globin gene cluster of a patient with δ,β-thalassemia (15) raises the possibility that recombination events mimicking the assembly of immunoglobulin and T-cell receptor genes may result in aberrant structural rearrangements of other unrelated genes (12,14).

The DNA sequences at the borders of the DNA inversion in the genome of the FC AI/CIII deficient patients were aligned with the corresponding sequences in the normal apoAI and apoCIII genes (Fig. 6A). This alignment shows that a hexanucleotide (AGACAG) is located near both breakpoints in the apoAI and apoCIII genes. In addition, three of the total five such hexanucleotides present in the 8.6 kb apoAI-apoCIII gene region are located near these breakpoints. However, there is no homology between this hexanucleotide and either consensus sequences of the hepta- or nona-nucleotides thought to serve as "joint signals" in the assembly of functional immunoglobulin and T-cell receptor genes (11,12,16,17). Interestingly, this hexanucleotide is present near the breakpoints of three out of six chromosomal translocations occurring within the c-myc proto-oncogene of certain mouse plasmacytomas (see M104E Sa, M603Sa and J558 Sa in Fig. 2C of ref 18). It is therefore possible that the DNA inversion in the genome of the FC AI/CIII deficient patients was

A 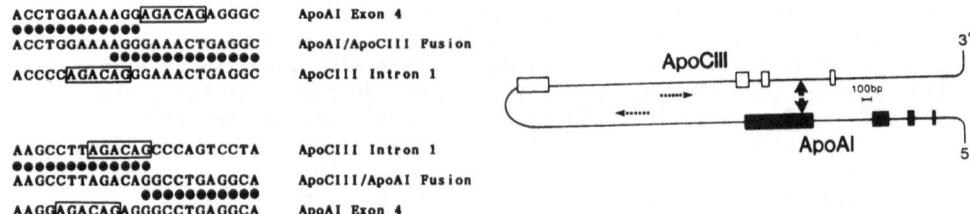 B

ACCTGGAAAAGG`AGACAG`AGGGC ApoAI Exon 4
●●●●●●●●●●●●●
ACCTGGAAAAGGGAAACTGAGGC ApoAI/ApoCIII Fusion
●●●●●●●●●●●●●●●●
ACCCC`AGACAG`GGAAACTGAGGC ApoCIII Intron 1

AAGCCTT`AGACAG`CCCAGTCCTA ApoCIII Intron 1
●●●●●●●●●●●●●
AAGCCTTAGACAGGCCTGAGGCA ApoCIII/ApoAI Fusion
●●●●●●●●●●●●
AAGG`AGACAG`AGGGCCTGAGGCA ApoAI Exon 4

Fig. 6.　(A) Comparison of nucleotide sequences located at either border
of the 6.0 kb DNA inversion. The hexanucleotides (AGACAG)
located near the break points of the DNA inversion are boxed.
Homologous sequences between the normal apoAI, apoCIII genes and
the break points of the DNA inversion are indicated by dots.
(B) Folding of the normal apoAI and apoCIII genes such that the
breakpoints (thick arrows) of the 6.0 kb DNA inversion are
aligned. The location of Alu-I repetitive DNA elements and
their relative orientation are indicated by dotted arrows. The
intron-exon organization of these genes is indicated as in Fig.
2.

generated by a recombinational mechanism subject to different structural
requirements from those necessary for the immunoglobulin and T-cell
receptor gene rearrangements, and that this mechanism may also be
involved in abortive immunoglobulin switching events resulting in
chromosomal translocations. Alternatively, it is possible that under
certain circumstances the immunoglobulin and T-cell receptor gene
recombinational system(s) may loosen its requirements for specific
"joining signals" and may catalyse recombination events between DNA
segments containing sequences other than those in the immunoglobulin and
T-cell receptor genes. In either case, however, the observed rarity of
recombination events leading to aberrant gene rearrangements does not
preclude the possibility that such events occur quite frequently, but do
not always result in identifiable phenotypes. It is also noteworthy that
AluI repetative DNA elements may be involved in facilitating
recombination events. For example, the possible involvement of Alu-I
repetitive DNA elements in generation of two different DNA deletions
within the low density lipoprotein (LDL) receptor gene of certain
patients with familial hypercholesterolemia (19,20), and in generation of
a deletion-inversion rearrangement within the β-globin gene cluster of a
patient with δ,β-thalassemia (15), suggests that these DNA elements may
stabilize DNA stem-loop structures necessary for recombination. The
apoAI-apoCIII gene cluster contains two Alu-I elements oriented in the
same direction (See Fig. 2). Interestingly these elements are located
within 1.5 kb from either breakpoint of the DNA inversion in the genome
of the FC AI/CIII deficient patients. Furthermore, folding of the apoAI
and apoCIII genes in a hairpin structure brings these Alu-I elements, as
well as the breakpoints of this DNA inversion in close physical proximity
(Fig. 6B). Therefore, a cross-over event within the stem of this hairpin
structure could result in the DNA inversion observed in the genome of
these patients. Is should also be pointed out that both phylogenetic
considerations of the evolution of the apoAI-CIII-AIV gene cluster (see
chapter　, page　, this book) and studies of the linkage disequilibrium
between restriction site polymorphisms of this gene cluster in several
different human populations (21) suggest that this cluster is involved in
recombination events. It may therefore be speculated that the DNA

inversion in the genome of the FC AI/CIII deficient patients resulted from such a recombination event but between misaligned nucleotide sequences.

Whatever the mechanism for the generation of the inversion in the genome of the FC AI/CIII deficient patients, this DNA rearrangement has resulted in aberrant expression of both apoAI and apoCIII genes. As a consequence the plasma of these patients is deficient in apoAI, apoCIII and HDL. That the HDL deficiency in these patients is due primarily to the inactivation of their apoAI gene is supported by the fact that apoAI is the major protein constituent of HDL and by recent data indicating that treatment of rats with β,β'-tetramethyl-substituted hexadecanedioic acid (MEDICA 16) results in a 10-fold reduction of apoCIII but it does not result in HDL deficiency (22). Therefore, HDL deficiency in these patients is caused by genetic deficiency in apoAI. The striking association of apoAI and HDL deficiency with the development of severe premature atherosclerosis in these patients provides direct evidence for a cause and effect relationship between diminished plasma apoAI and HDL levels and the development of this disease. In addition, this association provides strong presumptive evidence for an anti-atherogenic potential of apoAI and HDL.

Despite the overall agreement of the genetic (i.e. FC AI/CIII deficiency) and epidemiological studies with regard to the inverse relationship between plasma HDL levels and risk for atherosclerosis, there are certain inconsistencies. For example, the parents of the FC AI/CIII deficient patient, although they are heterozygotes for the aberrantly expressing apoAI and apoCIII genes and have abnormally low plasma HDL levels, do not present clinical evidence for atherosclerosis. Similar discrepancies have been noticed when epidemiological data obtained by in between-populations and within-population correlations of plasma HDL levels and atherosclerosis are compared. Since there is evidence that dietary cholesterol increases HDL levels, its has been proposed that the power of HDL plasma levels as an atherosclerosis risk factor reflects a negative correlation between susceptibility to atherosclerosis and the magnitude of diet-induced increment in HDL (3). Therefore, it is possible that the dietary habits and general lifestyles of the parents of these patients do not impose increased needs for elevation of their HDL plasma levels and that the normal apoAI gene present in one of their chromosomes provides sufficient apoAI and HDL to protect them against premature atherosclerosis. If this hypothesis is correct, the possible development of premature atherosclerosis in the offspring of these patients may be prevented by appropriate dietary adjustments.

REFERENCES

1. A.M. Gotto, H.J. Pownall and R.J. Havel, Introduction to the plasma lipoproteins, Methods Enzymol., 128:3 (1986).
2. M.S. Brown and J.L. Goldstein, Lipoprotein metabolism in the macrophage: implications for cholesterol deposition in atherosclerosis, Annu. Rev. Biochem., 52:223 (1983).
3. N.E. Miller, High density lipoportein as a predictor of clinical coronary heart disease, In "Atherosclerosis VII", N.H. Fidge, and P.J. Nestel eds., Elsevier Science Pub. p. 61 (1986).
4. R.A. Norum, J.B. Lakier, S. Goldstein, A. Angel, R.B. Goldberg, W.D. Block, D.K. Noffze, P.J. Dolphin, J. Edelglass, D.D. Bogorad and P. Alaupovic, Familial deficiency of apolipoproteins AI and CIII and precocious coronary-artery disease, N. Engl. J. Med. 306:1513 (1984).

5. S.K. Karathanasis, R.A. Norum, V.I. Zannis and J.L. Breslow, An inherited polymorphism in the human apolipoprotein AI gene locus related to the development of atherosclerosis, _Nature_ 301:718 (1983).

6. S.K. Karathanasis, E. Ferris and I.A. Haddad, DNA inversion within the apolipoproteins AI-CIII-AIV encoding gene cluster of certain patients with premature atherosclerosis, _Proc. Natl. Acad. Sci. USA_ 84:;7198 (1987)

7. S.K. Karathanasis, P. Oettgen, I.A. Haddad and S.E. Antonarakis, Structure, evolution and polymorphisms of the human apolipoprotein AIV gene (ApoA4), _Proc. Natl. Acad. Sci. USA_ 83:8457 (1986).

8. S.K. Karathanasis, V.I. Zannis and J.L. Breslow, Isolation and characterization of the human apolipoprotein AI gene, _Proc. Natl. Acad. Sci. USA_ 80:6147 (1983).

9. A.J. Davison and N.M. Wilkie, Inversion of the two segments of the herpes simplex virus genome in intertypic recombinants, _J. Gen. Virol._ 64:1 (1983).

10. R.H.A. Plasterk, R. Kanaar and P. VandePutte, A genetic switch in-vitro DNA inversion by gin protein of phage mu, _Proc. Natl. cad. Sci. USA_ 81:2689 (1984).

11. S. Lewis, A. Gifford and D. Baltimore, DNA elements are asymmetrically joined during the site-specific recombination of Kappa Immunoglobulin genes, _Science_ 228:677 (1985).

12. M. Malissen, C. McCoy, D. Blanc, J. Trucy, C. Devaux, A.-M. Schmitt-Verhulst, F. Fitch, L. Hood and B. Malissen, Direct evidence of chromosomal inversion during T-cell receptor β-gene rearrangements, _Nature_ 319:28 (1986).

13. Y. Tsujimoto, E. Jaffe, J. Cossman, J. Gorham, P.C. Nowell and C.M. Croce, Clustering of breakpoints on chromosome 11 in human B cell neoplasms with the 11; 14 chromosome translocation, _Nature_ 315:340 (1985).

14. J. Hochtl and H.G. Zachau, A novel type of aberrant recombination in immunoglobulin and genes and its implications for V-J joining mechanism, _Nature_ 302:260 (1983).

15. M.W. Jennings, R.W. Jones, W.G. Wood and D.J. Weatherall, Analysis of an inversion within the human beta globin gene cluster, _Nucleic Acids Res._ 13:2897 (1985).

16. J.G. Seidman and P. Leder, A mutant immunoglobulin light chain is formed by aberrant DNA splicing and RNA splicing events, _Nature_ 286:779 (1980).

17. P Leder, E.E. Max, J.G. Seidman, S.P. Kwan, M. Scharff, M. Nau and B. Norman, Recombintion events that activate diversify and delete immunoglobulin genes, Cold Spring Harbor Symposia on Quantitative Biology, Vol 45 Parts 1 & 2 Movable genetic elements; XXI+445P (Part 1) XIII+578P (Part 2) Cold Spring Harbor Laboratory, Cold Spring Harbor, NY USA p.859 (1981).

18. S.P. Picoli, P.G. Caimi and M.D. Cole, A conserved sequence at c-myc oncogene chromosomal translocation breakpoints in plasmacytomas, _Nature_ 310:327 (1984).

19. M.A. Lehrman, W.J. Schneider, T.C. Sudhof, M.S. Brown, J.L. Goldstein and D.W. Russell, Mutation in LDL receptor: Alu-Alu recombination deletes exons encoding transmembrane and cytoplasmic domains, _Science_ 227:140 (1985).

20. M.A. Lehrman, D.W. Russell, J.L. Goldstein and M.S. Brown, Exon-Alu recombination deletes 5 kilobases from the low density lipoprotein receptor gene, producing a null phenotype in familial hypercholesterolemia, _Proc. Natl. Acad. Sci. USA_ 83:3679 (1986).

21. S.E. Antonarakis, P. Oettgen, A. Chakravarti, S. Halloran, R.R. Hudson, L. Feisee and S.K. Karathanasis, DNA polymorphism haplotypes of the human apolipoprotein apoAI-apoCIII-apoAIV gene cluster, _Hum. Genet._ (In Press) (1988).

22. J. Bar-Tana, G. Rose-Kahn, B. Frenkel, Z. Shafer and M. Fainaru, Hypolipidemic effect of β,β'-methyl-substituted hexadecanedioic acid (MEDICA 16) in normal and nephrotic rats, <u>J. Lipid Res.</u> 29:431 (1988).

FAMILIAL APOLIPOPROTEIN A-I, C-III AND A-IV DEFICIENCY

Jose M. Ordovas, Dianne C. King and Ernst J. Schaefer

Lipid Metabolism Laboratory, USDA Human Nutrition

Research Center at Tufts University, Boston, U.S.A.

Apolipoprotein (apo) A-I is the major protein of high density lipoproteins (HDL), while apoC-III and apoA-IV are minor protein components which are also found within triglyceride-rich lipoprotein particles (1). Decreased levels of plasma HDL cholesterol and apoA-I have been demonstrated to be risk factors for premature coronary artery disease (CAD) (2,3). The genes coding for apoA-I, apoC-III, and apoA-IV are adjacent to one another on the long arm of chromosome 11 (4). Restriction fragment length polymorphisms (RFLPs) within this region (PstI site 3' to the apoA-I gene, and SstI within the 3' untranslated region of the apoC-III gene) have been associated with premature coronary atherosclerosis, decreased HDL, and hypertriglyceridemia (5-8). A kindred has been described in which two sisters developed severe coronary atherosclerosis in their late twenties and were noted to have marked HDL deficiency and trace or undetectable levels of apoA-I and apoC-III in their plasma. They also had diffuse yellow-orange planar xanthomas, and mild corneal opacification (9). Obligate heterozygotes in this kindred had levels of HDL cholesterol, apoA-I and apoC-III which were approximately 50% of normal. This disorder, known as familial apolipoprotein A-I and C-III deficiency, has recently been shown to be due to a DNA rearrangement involving the adjacent apoA-I and apoC-III genes (10) (more detailed description of this disorder is presented by Dr. S.K. Karathanasis in another chapter of the present volume).

We have previously described a similar kindred known as familial apolipoprotein A-I and C-III deficiency, variant II, in which the homozygous proband had an HDL cholesterol of 1 mg/dl, and undetectable levels of apoA-I and apoC-III as determined by radioimmunoassay (11,12). The female proband expired at age 45 secondary to diffuse coronary artery atherosclerosis documented at autopsy despite coronary artery bypass surgery. Her fasting plasma levels of very low density lipoprotein (VLDL) and low density lipoprotein (LDL) cholesterol levels were normal at 4 and 106 mg/dl respectively. She had no other known risk factors for atherosclerosis, and no evidence of liver disease. The patient had no xanthomas, but dis have mild corneal opacification. In addition she had alpha tocopherol levels that were

22% of normal, a slightly prolonged prothrombin time, and linoleic acid percentages within plasma phospholipids and cholesterol esters that were 54% and 72% of normal, indicative of possible moderate intestinal malabsorption (12). Obligate heterozygotes in this kindred had HDL cholesterol, apoA-I, and apoC-III levels that were approximately 60% of normal.

In order to define the molecular defect in this kindred subjected chromosomal DNA isolated from obligate heterozygotes to restriction endonuclease analysis utilizing PstI, SstI, BamHI, PvuII, HindIII, EcoRI and XmnI followed by Southern blotting analysis and hybridization with a PstI-PstI, 2.2 kb DNA fragment spanning the entire apoA-I gene. No RFLPs were detected with this probe and the enzymes previously described, or with probes spanning the apoC-III gene and the apo A-IV gene. However the hybridization signals observed in the autoradiograms for heterozygotes were consistently weaker than the signals observed for controls. Analysis of gene copy number as previously described (13) confirmed that with the probes used, only one allele was present in this region. When a probe spanning a 1.1 kb region 3 kb 5' to the apoA-I gene was used, multiple RFLPs were detected with all the enzymes utilized, indicating a dramatic change in the structure of the DNA in this region. Utilizing this methodology, we were able to demonstrate the presence of two alleles. These data and sequencing analysis indicate that the defect in this kindred is due to a deletion of the entire apoA-I, apoC-III and apoA-IV gene locus.

Twenty one members of this kindred were analyzed and ten were found to be heterozygotes for this disorder. The heterozygotes had apolipoprotein A-I, C-III and A-IV values, determined by immunoassay that were 57%, 45% and 56% respectively of values obtained for unaffected relatives. HDL cholesterol values were also significantly lower (60%) in affected individuals than in controls. No significant differences were found for plasma apoB or LDL-cholesterol.

This data indicate that the defect in this kindred is due to the deletion of the apoA-I, apoC-III, and apoA-IV gene cluster, and supports the concept of a role for apoA-I deficiency in the pathogenesis of atherosclerosis. Moreover, the absence of apoA-IV in this kindred points to a role for apoA-IV in the intestinal absorption of alpha-tocopherol, vitamin K and essential fatty acids, since homozygotes for the previously reported apoA-I and apoC-III deficiency (9) did not have deficiencis of these constituents.

ACKNOWLEDGMENTS

Supported by grant HL35243 from the National Institutes of Health and contract 53-3K06-5-10 from the U.S. Department of Agriculture Research Service.

REFERENCES

1. A.J. Scanu, J. Toth, C. Edelstein and E. Stiller. Fractionation of human serum high density lipoproteins in urea solutions; evidence of polypeptide heterogeneity. Biochemistry 8:3309-3316 (1969).

2. G. J. Miller, and N. F. Miller, Plasma high density lipo-
    protein concentration and development of ischemic heart
    disease. Lancet 1:16-20 (1975).

3. J. J. Maciejko, D. R. Holmes, B. A. Kottke, A. R. Zinsmeis-
    ter, D. M. Dinh and S. J. T. Mao, Apolipoprotein A-I as
    a marker of angiographically assesed coronary artery disea-
    se. N. Engl. J. Med. 309:385-389 (1983).

4. S. K. Karathanasis, Apolipoprotein multigene family:
    tandem organization of human apolipoprotein A-I, C-III and
    A-IV genes. Proc. Natl. Acad. Sci. U.S.A. 82:6374-6378
    (1985).

5. J. M. Ordovas, E. J. Schaefer, D. Salem, R. Ward, C. H.
    Glueck, C. Vergani, P.W.F. Wilson and S. K. Karathanasis,
    Apolipoprotein A-I gene polymorphism associated with pre-
    mature coronary artery disease and familial hypoalphali-
    poproteinemia. N. Engl. J. Med. 314:671-677 (1986).

6. A. Sidoli, G. Giudici, M. Soria and C. Vergani, Restric-
    tion fragment length polymorphisms in the A-I-C-III gene
    complex ocurring in a family with hypoalphalipoproteinemia.
    Atherosclerosis 62:81-87 (1986).

7. R. A. Anderson, T. J. Benda, R. B. Wallace, S. L. Eliason,
    J. Lee and T. L. Burns, Prevalence and associations of
    apolipoprotein A-I-linked DNA polymorphisms: results from
    a population study. Genet. Epidemiol. 3:385-397 (1986).

8. A. Rees, C. C. Shoulders, J. Stocks, D. J. Galton and
    F. E. Baralle, DNA polymorphism adjacent to human apoli-
    poprotein A-I gene: relation to hypertrigliceridemia.
    Lancet 1: 444-446 (1983).

9. R. A. Norum, J. B. Lakier, S. Goldstein, A. Angel, R. B.
    Goldberg, W. D. Black, D. K. Noffze, P. J. Dolphin, J.
    Edelglass, D. D. Borograd and P. Alaupovic, Familial
    deficiency of apolipoproteins A-I, C-III and precocious
    coronary artery disease. N. Engl. J. Med. 306:1513-1519
    (1983).

10. S. K. Karathanasis, E. Ferris, I. A. Haddad, DNA inver-
    sion within the apolipoproteins AI/CIII/AIV encoding gene
    cluster of certain patients with premature atherosclero-
    sis. Proc. Natl. Acad. Sci. U.S.A. 84:7198-7202 (1987).

11. E. J. Schaefer, W. H. Heaton, M. G. Wetzel and H. B. Bre-
    wer Jr. Plasma apolipoprotein A-I absence associted with
    a marked reduction of high density lipoproteins and pre-
    mature coronary artery disease. Arteriosclerosis 2:16-26
    (1982).

12. E. J. Schaefer, J. M. Ordovas, S. Law, G. Ghiselli, M. L.
    Kashyap, L. S. Srivastava, W. H. Heaton, J. J. Albers,
    W. E. Conners and H. B. Brewer Jr. Familial apolipoprotein
    A-I and C-III deficiency, variant II, J. Lip. Res. 26:1089-
    1101 (1985).

13. R. E. Corin, T. Turner, and P. Szabo. Murine Erythroleu-
    kemia cell variants: isolation of cells that have amplified
    the Dihydrofolate Reductase gene and retained the ability
    to be induced to differentiate. Biochemistry 25:3768-
    3773 (1986).

PHENOTYPIC EXPRESSION OF HEPATIC LIPASE DEFICIENCY

Philip Connelly, Graham McGuire
Maureen Lee, Ruth McPherson
and Alick Little

Lipid Research Clinic,
St. Michael's Hospital
30 Bond Street,
Toronto, Canada M5B 1W8

INTRODUCTION

Lipolytic processes are essential for the delivery of lipoprotein derived
fatty acids to peripheral tissues. Lipolysis also plays a fundamental
role in the modification of lipoprotein structure and apolipoprotein
epitope expression with ultimate effects on the cellular uptake of
apolipoproteins and lipids. Two enzymes, lipoprotein lipase (LPL) and
hepatic lipase (HL) account for the majority of triglyceride hydrolysis in
human plasma (1). Both can be released into the circulation after the
intravenous injection of heparin. The role of LPL has been studied
extensively and this enzyme, in the presence of apolipoprotein CII
(apo CII), hydrolyzes triglycerides in very low density lipoprotein (VLDL)
and chylomicrons, initiating conversion to intermediate density lipoproteins
(IDL) and chylomicron remnants (CR). This process is accompanied by
transfer of apo C, apolipoprotein A (apo A) phospholipid (PL) and free
cholesterol (FC) to HDL (2). LPL is thus an important determinant of both
HDL and triglyceride levels and there is a clear positive association
between LPL activity and triglyceride concentrations in normal and
hypertriglyceridemic subjects and a reduction of HDL cholesterol in LPL
deficiency.

FUNCTION OF HEPATIC LIPASE

The physiological role of HL has been less clearly defined. This is a
phospholipase and triglyceride lipase located at the luminal surface of
hepatic nonparenchymal cells which functions in the remodelling of HDL
and apo B containing lipoproteins (3). _In vitro_ studies have demonstrated
that HL hydrolyzes HDL triglyceride and phospholipid. IDL and the less
dense subfraction of LDL (1.019-1.045) are also substrates for HL activity
(4,5). Similarly, _in vivo_ inhibition of HL by intravenous infusion of
anti-HL antibody into rats and cynomolgous monkeys resulted in increases
in HDL and IDL triglyceride and phospholipid and in VLDL triglyceride (6,7)

Population studies have demonstrated an inverse relationship between HL activity and HDL2 cholesterol levels and a positive correlation between HL and VLDL triglyceride (8,9). Certain pharmacological agents such as anabolic steroids (stanozolol) (10) and progestins (norgestral) (11) increase HL activity and this increase is accompanied by a reduction in HDL2 and an elevation in LDL cholesterol concentrations. These studies demonstrate that specific detrimental effects on lipoprotein structure can be identified in situations in which hepatic lipase activity is increased. A relative deficiency of hepatic lipase may also produce abnormalities in the structure and function of lipoproteins which may alter risk for atherosclerotic disease.

PLASMA LIPOPROTEINS IN HEPATIC LIPASE DEFICIENCY

This laboratory as identified a pedigree with HL deficiency (12). The phenotypic characteristics of these subjects included hypertrigly-ceridemia, beta migrating VLDL, and triglyceride and phospholipid enrichment of HDL and LDL. The activity of postheparin LPL was normal but postheparin HL activity was absent. Recently another family with virtually identical lipoprotein abnormalities has been described by Carlson (13).

The St. Michael's pedigree has premature coronary heart disease and elevated apo B levels in contrast to those described by Carlson who have normal apo B and no readily apparent manifestations of athero-sclerosis. Further investigations of the family described at this clinic have revealed that elevated levels of VLDL and LDL occur in affected and nonaffected family members suggesting the presence of additional genes for hyperlipoproteinemia in this family independent of hepatic lipase deficiency (14).

The cholesterol and triglyceride composition of the d 1.006 - 1.063 g/ml plasma lipoproteins of four family members with hepatic lipase deficiency and the 16 unaffected family members is illustrated in Figure 1. There was a very marked increase in the triglyceride content of the LDL and IDL in affected subjects as compared to family members with normal HL activity (14). (Figure 1).

Similarly, HDL isolated from HL deficient subjects demonstrated significant triglyceride enrichment (14). (Figure 2). These obser-vations are consistent with the known function of HL as a triglyceride lipase and phospholipid lipase.

A further significant abnormality in hepatic lipase deficiency is the presence of beta-migrating VLDL. In Type III hyperlipoproteinemia, B-VLDL derived from hepatic VLDL or intestinal chylomicrons accumulates because of a defect in the removal of CR, and VLDL remnants due to the presence of abnormal (E2) or dysfunctional E isoforms (15). B-VLDL is well established as an atherogenic lipoprotein (16).

The B-VLDL which accumulates in HL deficiency differs from type III B-VLDL in that the E isoform is E3/E3 and little apo B48 is present (17). Further studies are in progress to determine the interaction of these particles with the fibroblast B/E receptor and on cellular accumulation of cholesterol.

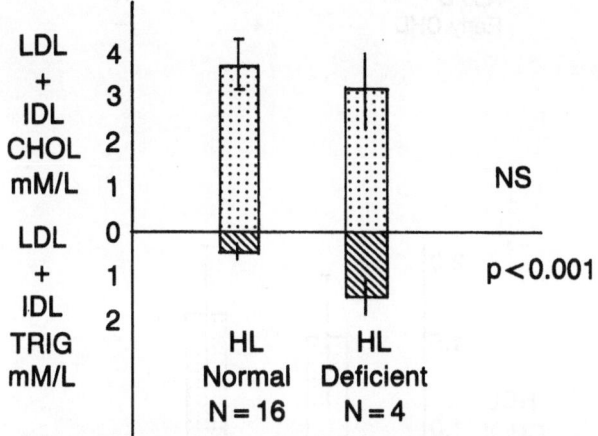

FIGURE 1 : Lipid composition of apo B containing lipoproteins in affected and nonaffected members of hepatic lipase deficient pedigree (14).

TABLE 1   Phenotypic characteristics of patients with hepatic
          lipase deficiency : Comparison of two pedigrees (12,13).

**Phenotypic Characteristics
of Hepatic Lipase Deficiency**

	Toronto	Stockholm
VLDL - TRIG	↑	↑
LDL - TRIG/PL	↑	↑
HDL - TRIG/PL	↑	↑
β Migrating VLDL	+	+
Apo B	↑	N
Early CHD	+	−

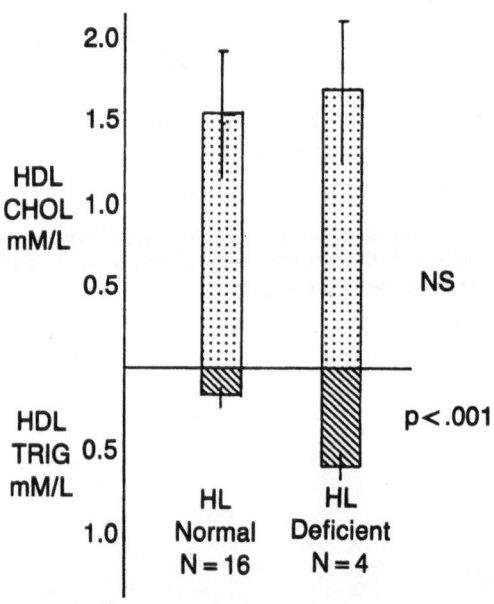

FIGURE 2 : Lipid composition of HDL in affected and nonaffected
           members of hepatic lipase deficient pedigree (14).

REFERENCES

1. P. Nihlsson-Ehle, A.S. Garfinkel, M.C. Schotz, Annu Rev Biochem. 49:667-693 (1981).
2. R.J. Havel, J.P. Kane, M.L. Kashyap, J Clin Invest. 52:32-38 (1973).
3. R.L. Jackson, in The Enzymes, P. Boyer, ed, Academic Press, New York, N.Y., 16:141-181 (1983).
4. J.A. Little, P.W. Connelly, Adv Exp Mol Med. 201:253-260 (1985).
5. Y. Homma, N. Nakaya, H.Nakamura, Y. Goto, Artery. 13:19-31 (1985).
6. J. Grosser, V. Schrecker, H. Greten, J Lipid Res. 22:437-442 (1981).
7. I.J. Goldberg, N.A. Le, J.R. Paterniti, J Clin Invest. 70:1184-1192 (1982).
8. T. Kuusi, P. Saarinen, E.A. Nikkila, Atherosclerosis 36:5879-593 (1980).
9. D. Applebaum-Bowden, S.M. Haffner, P.W. Wahl, Arteriosclerosis 5:273-282 (1985).
10. D. Applebaum-Bowden, S.M. Haffner, W.R. Hazzard, Metabolism 36:949-952 (1987).
11. M.J. Tikkanen, E.A. Nikkila, T. Kuusi, S. Sipenen, Clin Chim Acta 115:63-69 (1981).
12. W.C. Breckenridge, J.A. Little, P. Alaupovic, Atherosclerosis 45:161-169 (1982).
13. L.A. Carlson, L. Holmquist, D. Nihlsson-Ehle, Acta Med Scand. 219:435-447 (1986).
14. P.W. Connelly, M. Lee, G.F.McGuire, J.A. Little, (submitted to Arteriosclerosis, 1988).
15. V.I. Zannis, J.I. Breslow, Adv Human Genetics. 14:125-148 (1985).
16. R.W. Mahley, T.L. Innerarity, M.S. Brown, Y.K. Ho, J.L. Goldstein, J Lipid Res. 21:970-980 (1980).
17. P.W. Connelly, S. Ranganathan, G.F. McGuire, M.Lee, J.J. Myher, B.A. Kottke and J.A. Little, J Biol Chem. In press.

ABNORMAL PROCESSING OF HDL PRECURSORS IN TANGIER MONOCYTE DERIVED MACROPHAGES

G.Schmitz, H.Robenek[*], B.Brennhausen, G. Assmann

Institut für Klin. Chemie und Laboratoriumsmedizin
der Universität Münster und Arbeitsgruppe Zell-
biologie[*], Albert-Schweitzer-Str. 33
4400 Münster, FRG

Tangier disease is characterized by the absence of normal HDL
in plasma and by the accumulation of cholesteryl esters in
various tissues. These include the liver, spleen, lymph
nodes, thymus, intestinal mucosa, skin and proably the
cornea.
The combination of HDL-deficiency with a low plasma
cholesterol concentration accompanied by normal or elevated
plasma triglycerides and enlarged orange-yellow tonsils and
splenomegaly is pathognomonic for the disease (1).
The major apolipoproteins of normal HDL are apolipoprotein
A-I (apo A-I) and apolipoprotein A-II (apo A-II). In Tangier
subjects, serum concentrations of these apolipoproteins are
< 1 % and 5-10 % of normal, respectively. In spite of the
diminished concentration of apo A-I in serum, small
intestinal epithelial cells from Tangier patients contain
normal amounts of apo A-I. Metabolic studies have indicated
that the decreased concentration of apo A-I in Tangier
patients is due to the enhanced catabolism of this
apolipoprotein. Detailed studies have established that there
is no defect in the structure of the propeptide or the
converting peptidase and that the rate of conversion of
proapo A-I to mature apo A-I in Tangier disease is similar to
normal. Moreover, studies of the cDNA derived amino acid
sequence of Tangier apo A-I have clearly shown that apo A-I
from Tangier patients was identical to normal and this was
also shown for apo A-II. Thus the structure of both apo A-I
and and apo A-II is normal in Tangier disease. Therefore it
seems likely that the primary abnormality in Tangier disease
involves increased catabolism of the HDL-constituents at the
cellular level. Hypercatabolism of HDL is best explained by
a defect in the interaction of HDL with cells critically
dependent upon HDL-mediated cholesterol efflux.
Macrophage storage of cholesteryl esters is the outstanding
pathologic finding in Tangier disease and it seems most
likely that HDL-hypercatabolism is associated with these
cells.
The mechanism of cholesterol accumulation in macrophages
has been intensively investigated in recent years and it has

been demonstrated that the uptake of lipoprotein bound
cholesterol in macrophages is mediated predominantly by
receptor-dependent endocytosis (2,3). However, the regulation
of cholesterol homeostasis in macrophages is clearly
different from that of apo B,E-receptor (LDL-receptor) cells
such as fibroblasts. In the apo B,E-receptor cells, the
activity of the B,E-receptor and the rate limiting step of
intracellular cholesterol synthesis (HMG-CoA-reductase)
activity are strictly controlled. Therefore, these cells do
not need potent cholesterol acceptors for the secretion of
surplus cholesterol.

Macrophages, however, are highly dependent on effective
cholesterol acceptors in their vicinity to prevent
transformation into foam cells when challenged by an
increased cholesterol uptake. When macrophages ingest
cholesterol, e. g. membranes, cells debris or atherogenic
lipoproteins, the absorbed cholesteryl esters are hydrolzed
(Figure 1) by lysosomal acid cholesteryl ester hydrolase
(ACEH), and cholesterol is released into the cytoplasm and
reesterified to form cytoplasmic cholesteryl ester-containing
droplets, which in turn are hydrolyzed and reesterified in a
continuous cycle (2,3).

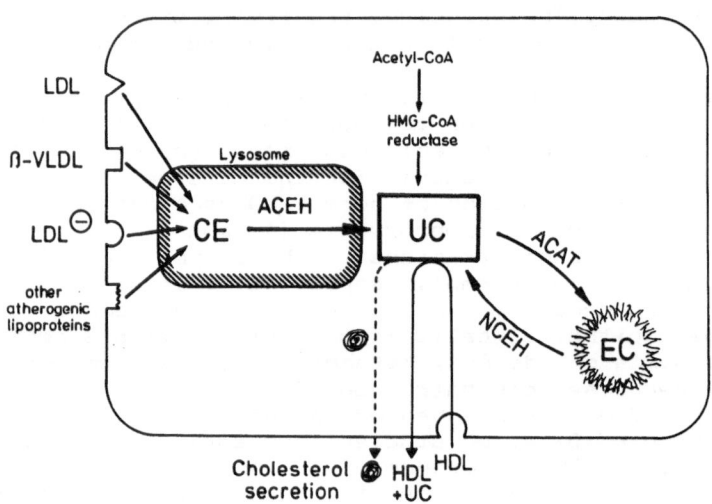

Figure 1. Cholesterol metabolism in macrophages

Hydrolysis of cholesteryl esters in the cytoplasmic
compartment is mediated by neutral cholesteryl ester
hydrolase (NCEH) which is probably located at the margin of
the cytoplasmic lipid droplets. Esterification is mediated by
microsomal acyl-CoA:cholesterol acyltransferase (ACAT)
reaction, which esterifies cholesterol with a fatty acid
(mainly oleic acid).

If the surrounding medium contains a cholesterol acceptor
such as high density lipoproteins (HDL), the cholesterol is
not reesterified or stored but mainly released as
unesterified cholesterol.

It has been demonstrated that HDL absorbs unesterified

cholesterol from cell surface membranes and it has been
hypothesized that this leads to an imbalance between
cytoplasmic cholesterol and the cholesterol content of the
cell surface which is assumed to be reequilibrated from the
cytoplasmic cholesterol pool. HDL are seen to be effective
acceptors for cholesterol since the outer surface of the HDL
particle has a low cholesterol/phosphatidylcholine ratio in
comparison to most of the cell membranes.
Using biochemical and morphological methods we have recently
shown that apo-AI-containing HDL bind to specific surface
receptors on macrophages (4-6) and apolipoprotein A-I, the
major HDL apolipoprotein, might be the ligand of this
receptor (Figure 2).

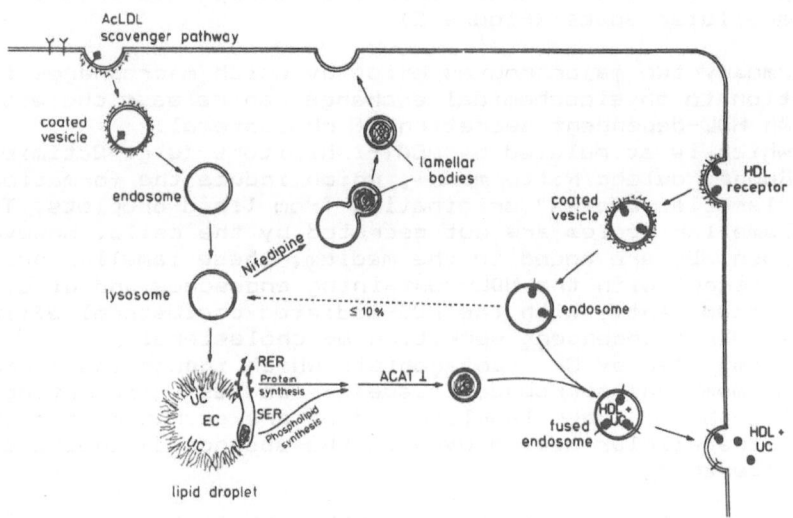

Figure 2. Cholesterol release from macrophages

HDL bind to this surface receptor, are internalized in a non-
lysosomal compartment and upon uptake of free cholesterol are
released from the cells as native particles. HDL take up
cholesterol from cholesterol rich "lamellar bodies" which
originate from cytoplasmic lipid droplets. These "lamellar
bodies" are formed upon attachment of endoplasmic reticulum
to the margin of the lipid droplets, and are surrounded by a
newly synthesized membrane.
This HDL mediated cholesterol efflux is promoted by ACAT-
inhibitors (5) which increase cellular free cholesterol.
The ligand specificity of the purified 110 kDa HDL-receptor
protein was analyzed and revealed a higher affinity for
discoidal HDL precursors and small spherical apo A-I-
containing HDL than for mature HDL particles.
In addition to the HDL mediated cholesterol efflux from
macrophages we have investigated in great detail a second
major pathway (Figure 2), which can be amplified by "slow
calcium channel" blockers of the dihydropyridine type (7,8).
Ca^{++}-antagonists have been recently shown in various studies
to exert antiatherogenic effects in vivo and to decrease in
vitro the cholesterol content of several cells including
smooth muscle cells and monocyte/macrophages. This reduction
of cellular cholesterol is, however, not associated with an

elevated level of HDL-binding activity (7). Therefore, other mechanisms independent from the HDL-receptor pathway must account for this effect.

When macrophages are incubated with acetyl-LDL in the presence of a Ca^{++}-antagonist, profound morphological and biochemical changes can be observed. The foamy organelles which represent cholesterol loaded endosomes and lysosomes are transformed to lamellar structures upon incubation, which ultimately leads to fragmentation of these foamy organelles into small "lamellar bodies". Each lamellar body is surrounded by a membrane which may be derived from the original lysosomal membrane. Upon prolonged exposure to acetyl-LDL and the Ca^{++}-antagonist, these lamellar bodies become increasingly condensed, move towards the cell periphery and are released from the macrophages into the extracellular space (Figure 2).

In summary two major routes exist by which macrophages in addition to physicochemical exchange can release cholesterol:
1. An HDL-dependent secretion of cholesterol,
    which is stimulated by ACAT-inhibitors (e.g. Octimibate, Rhone-Poulenc/Nattermann), which induce the formation of "lamellar bodies" originating from lipid droplets. These lamellar bodies are not secreted by the cells. However, when HDL are added to the medium, these lamellar bodies interact with the HDL-containing endosomes and disappear concomitantly with the HDL-mediated cholesterol efflux.
2. An HDL-independent secretion of cholesterol,
    stimulated by Ca^{++}-antagonists which induce the formation of membrane surrounded "lamellar bodies" originating from lysosomes. These lamellar bodies are secreted into the extracellular medium even in the absence of cholesterol acceptors.

The interaction of normal human HDL with isolated human monocytes has been studied in Tangier patients and controls (1,9,10). It was observed that normal human monocytes, similar to mouse peritoneal macrophages bind apo A-I containing HDL to the cell surface receptor, internalize the bound HDL particles and transport the internalized HDL through the cytoplasmic compartment without significant lysosomal degradation (Figure 3a-d). Ultimately, HDL are resecreted from the cells. Treatment of Tangier monocyte-derived macrophages with normal HDL, however, resulted in a slight increase in HDL binding, a failure of resecretion of internalized HDL and a trapping of these lipoproteins in the lysosomal compartment (Figure 3e) followed by lysosomal degradation (Figure 4).

We have concluded from these experiments that the molecular defect in Tangier disease does not likely reside in either abnormal HDL apolipoproteins or abnormal HDL receptor recognition but could rather lie in intracellular events involved in the assembly of cellular cholesterol with the internalized HDL and its transport back to the cell surface. Such transcellular channelling might be impaired in Tangier disease because of functional abnormalities in the HDL-receptor, alterations in endosomal traffic normally preventing degradation of HDL by lysosomes, or a defect in a Golgi associated processing mechanism affecting assembly of cellular cholesterol with HDL or its subsequent resecretion.

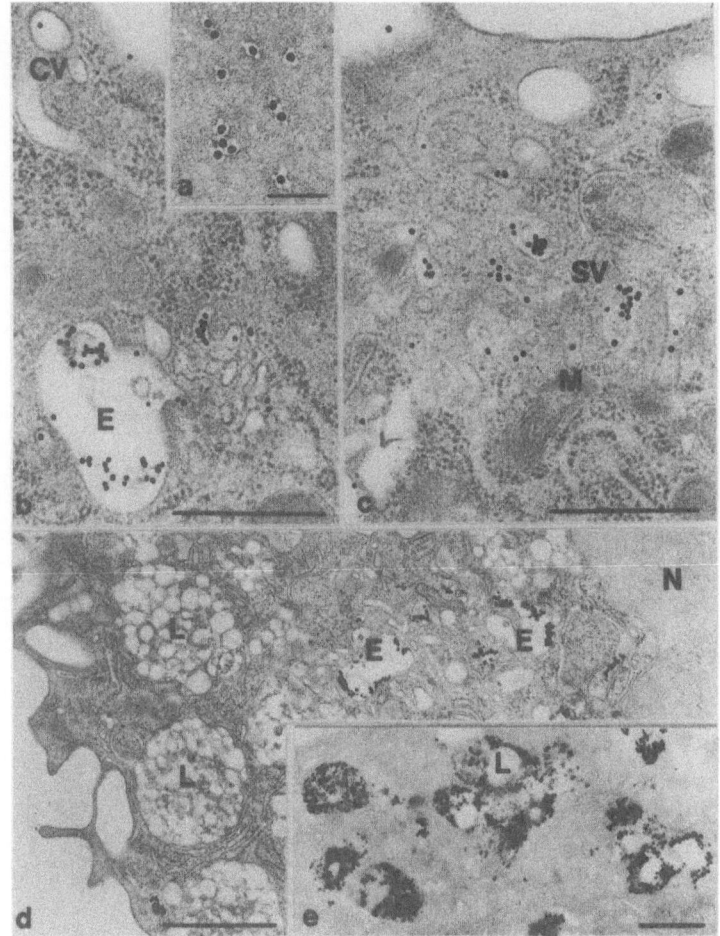

Figure 3. Uptake of HDL in peritoneal macrophages (a-d) and
        monocyte-derived macrophage of a Tangier patient
        (e).
a) HDL-gold conjugate.
b) Coated vesicle (CV) and endosome (E), labeled with
    HDL-gold conjugate.
c) Smooth vesicles (SV) labeled with HDL-gold
    conjugate. M= Mitochondria
d) Macrophage showing gold-labeled HDL internalized in
    endosomes (E).
    N= Nucleus, L= Lysosomes
e) Monocyte-derived macrophage of a Tangier patient
    incubated with gold-labeled HDL. HDL-gold complexes
    are found in acid phosphatase-positive vesicles
    which are regarded to represent secondary lysosomes
    (L).

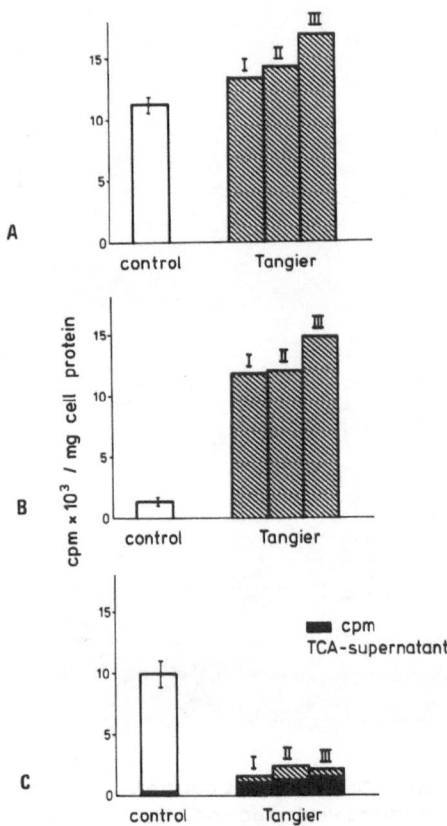

Figure 4. Binding, uptake and secretion of normal ^{125}I-HDL by
cultured blood monocytes from Tangier patients
(I-III) and controls.

All of these postulated defects could explain lysosomal
degradation of HDL and HDL precursors and thus we have
defined Tangier disease as a disorder of intracellular
organelle traffic (Figure 5).

Despite the abnormalities in HDL metabolism in these patients
the amount of clinically evident premature vascular disease
in homozygous Tangier patients is not striking. Therefore
with respect to the epidemological data concerning the strong
relation between HDL levels and the risk for CHD the question
arises why these patients do not suffer from premature
atherosclerosis. The reduced LDL concentration and the
substitution of triglycerides for cholesteryl esters as the

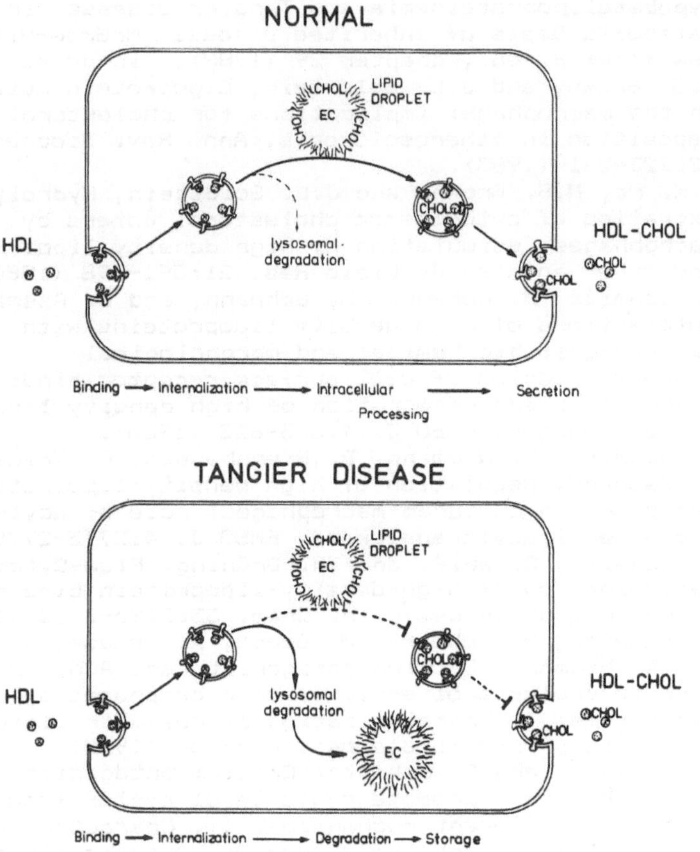

NORMAL

TANGIER DISEASE

Figure 5. Tangier disease: A disorder of intracellular
organelle traffic.

major neutral lipid in Tangier-LDL may lead to the observed
reduction in cholesteryl ester hydrolysis in the aortic wall
of these patients which may be an important factor which
protects these patients. Moreover, the thrombocytopenia
and hyporeactivity and storage pool deficiency of platelets
might significantly diminish the susceptibility to premature
atherosclerosis. In addition, our observation that there
exist a second route for an HDL independent secretion of
cholesterol containing "lamellar bodies" (Figure 2) might
enable the affected Tangier cells to secrete cholesterol

from these cellular cholesterol pools which critically
influence the atherogenic process and thereby accumulate
cholesteryl esters only in the metabolically inert
cytoplasmic lipid droplets.
However, further studies are necessary to evaluate the
different hypotheses and it may come out that all of the
factors listed above have a protective effect which prevent
early atherosclerosis in Tangier disease.

References

1. G. Assmann, G. Schmitz, and H.B. Brewer, Familial
    lipoprotein deficiency: Abetalipoproteinemia,
    Hypobetalipoproteinemia and Tangier Disease. in: The
    Metabolic Basis of Inherited Disease. McGraw-Hill,
    New York; 6. ed., chapter 29 (1988), (in press).
2. M.S. Brown, and J.L. Goldstein, Lipoprotein metabolism
    in the macrophage: implications for cholesterol
    deposition in atherosclerosis, Ann. Rev. Biochem.
    52:223-261 (1983).
3. Y.K. Ho, M.S. Brown, and J.L. Goldstein, Hydrolysis and
    excretion of cytoplasmic cholesterol esters by
    macrophages: stimulation by high density lipoproteins
    and other agents, J. Lipid Res. 21:391-398 (1980).
4. G. Schmitz, H. Robenek, H. Lohmann, and G. Assmann,
    Interactions of high density lipoproteins with
    macrophages: biochemical and morphological
    characterization of cell surface receptor binding,
    endocytosis and resecretion of high density lipoproteins
    by macrophages, EMBO J. 4:613-622 (1985).
5. G. Schmitz, R. Niemann, B. Brennhausen, R. Krause, and
    G. Assmann, Regulation of high density lipoprotein
    receptors in cultured macrophages: role of acyl-CoA:
    cholesterol acyltransferase, EMBO J. 4:2773-2779 (1985).
6. G. Schmitz, G. Wulf, and Th. Brüning, Flow-Cytometric
    Determination of high-density-lipoprotein binding sites
    on human leukocytes, Clin. Chem. 33:2195-2203 (1987).
7. G. Schmitz, H. Robenek, M. Beuck, R. Krause, A. Schurek,
    and R. Niemann, Caltium antagonists and ACAT-inhibitors
    promote cholesterol efflux from macrophages by different
    mechanisms. I. Charaterization of cellular lipid
    metabolism, Arteriosclerosis 8:46-56 (1988).
8. H. Robenek, and G. Schmitz, Calcium antagonists and
    ACAT-inhibitors promote cholesterol efflux from macro-
    phages by different mechanisms. II. Characterization of
    intracellular morphologic changes, Arteriosclerosis 8:
    57-67 (1988).
9. G. Schmitz, G. Assmann, H. Robenek, and B. Brennhausen,
    Tangier disease: A disorder of intracellular membrane
    traffic, Proc. Natl. Acad. Sci. USA, 82:6305-6309
    (1985).
10. G. Schmitz, G. Assmann, B. Brennhausen, and H.-E.
    Schaefer, Interaction of Tangier lipoproteins with
    cholesteryl ester-laden mouse peritoneal macrophages,
    J. Lipid Res. 28:87-99 (1987).

APO E POLYMORPHISM IN RELATION TO THE EXPRESSION OF

FAMILIAL DYSBETALIPOPROTEINEMIA

Louis M. Havekes[1], Peter de Knijff[1], Jan Gevers Leuven[1] and
Rune R. Frants[2]

[1]Gaubius Institute TNO, Herenstraat 5d, 2313 AD Leiden, and
[2]Dept. of Human Genetics, State University of Leiden,
Wassenaarseweg 72, 2333 AL Leiden, The Netherlands

INTRODUCTION

The apolipoprotein E (Apo E) present on chylomicron and very low
density lipoprotein (VLDL) remnants plays a central role in the hepatic
metabolism of these particles as this apolipoprotein is recognized with high
affinity by hepatic lipoprotein receptors (Sherrill et al., 1980; Weisgraber
et al., 1982). Human Apo E can be separated by isoelectric focusing into
three major isoforms, i.e. E2, E3 and E4 which differ in pI by a single
charge unit, Apo E4 being the most basic and E2 the most acidic form. This
heterogeneity of Apo E is the result of three different alleles, E*4, E*3
and E*2 at a single APOE locus in the long arm of chromosome 19 (Zannis and
Breslow, 1981; Utermann et al., 1982).

Apo E3 is the most commonly occurring or wild type form. Apo E4 is
thought to be derived from E3 by a Cys $\rightarrow$ Arg substitution at position 112
and is designated as E4 (Cys 112 $\rightarrow$ Arg). The most frequent Apo E2 is derived
from E3 by an Arg $\rightarrow$ Cys substitution at position 158 and is designated as E2
(Arg 158 $\rightarrow$ Cys). At present, a number of very rare mutants of Apo E have
been described. Some of these variants are more basic than Apo E4 or more
acidic than Apo E2, while other variants have the same electric charge as E2
or E3 (Rall et al., 1982; Innerarity et al., 1984; Yamamura et al., 1984,
Havel et al., 1983; Ghiselli et al., 1984; Havekes et al., 1986).

In contrast to E3 and E4, Apo E2 and most of the rare variants of Apo E
exhibit a reduced activity for binding to lipoprotein receptors which
results in the clinical picture of dysbetalipoproteinemia. Dysbetalipoproteinemia (FD, type III hyperlipoproteinemia according to Fredrickson et
al. (1980)) is the result of the accumulation of chylomicron and VLDL
remnants (Schneider et al., 1981). Most of the FD patients are homozygous
E2/E2. However, only about 4% of the E2/E2 homozygotes develop FD (Fredrickson et al., 1980).

FD patients with only heterozygosity for the E*2 allele are very rare.

E2/E2 homozygosity and the expression of familial dysbetalipoproteinemia (FD)

The reduced penetrance (4%) of FD in E2/E2 homozygotes is at least
partly due to the requirement of additional genetic or long lasting environmental factors. Moreover, a doctors delay is likely to play a role in
estimating the prevalence of the disease, as the symptoms of FD, xantho-

chromia striata palmaris and tuberous xanthomas on the elbows, are rare and not readily recognized.

To investigate which part of the E2/E2 homozygotes from the general population really develop FD, a relatively large random population sample was screened for apo E phenotype and plasma lipoprotein levels as only 1% of the general population exhibits the E2/E2 phenotype. In this study we determined the apo E phenotype distribution in 2,000 35-year old males randomly selected from the Dutch population (Klasen et al., 1987). This screening for Apo E phenotype was performed by isoelectric focusing of delipidated plasma (1-5 μl) followed by immunoblotting (Havekes et al., 1987). An example of this Western blot method is shown in Fig. 1. The Apo E phenotype distribution is presented in Table 1.

From these results it appeared that 13 out of 2,018 randomly selected 35-year old males exhibit the E2/E2 phenotype. All these E2/E2 homozygotes were analysed for lipoprotein pattern.

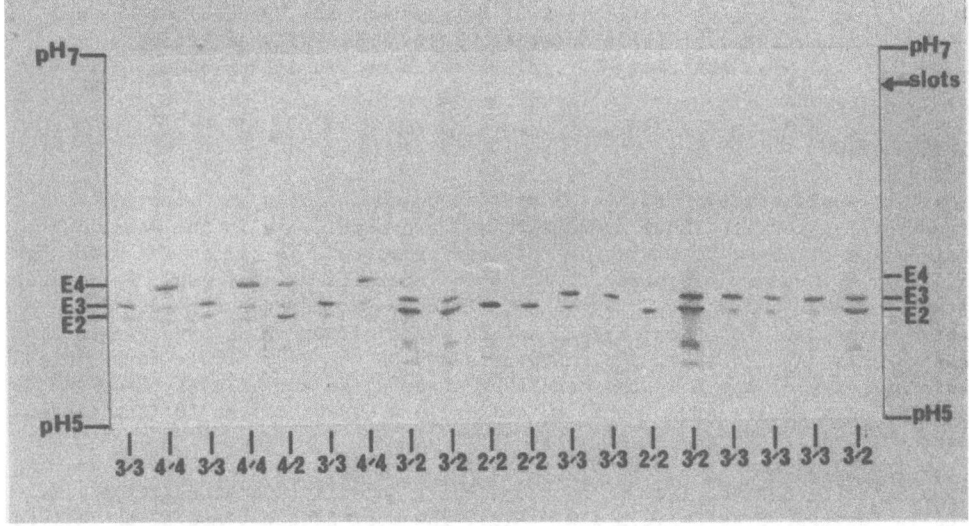

Fig. 1. Immunoblot of an isoelectric focusing slab gel (pH range 5-7) showing the Apo E phenotypes of 20 normolipidemic individuals.

Table 1. Apo E phenotype distribution in a random population sample of 2,018 35-year old males.

Phenotype	Number	Relative frequency
E4/E4	59	2.9
E4/E3	512	25.4
E4/E2	45	2.2
E3/E3	1,128	55.9
E3/E2	261	12.9
E2/E2	13	0.7
total	2,018	100

Lipid analyses and agarose electrophoresis showed that three out of the 13 E2/E2 homozygotes appeared to exhibit a typical FD lipoprotein pattern. Two additional E2/E2 homozygotes clearly showed the presence of d < 1.006 g/ml lipoproteins with β electrophoretic mobility, but without hyperlipoproteinemia. In contrast with typical FD patients none of these E2/E2 homozygotes exhibit tuberous xanthomas on the elbows and/or xanthochromia striata palmaris. These results show that as much as 23 percent of the male E2/E2 homozygotes randomly selected from the general population develop FD-like hyperlipoproteinemia at the age of 35 years. Although a population sample of 13 E2/E2 homozygotes is certainly too small for statistical analyses, we would like to point out that the percentage of the E2/E2 homozygotes that develope FD is higher than the 1 to 4 percent commonly suggested in the literature. The reason for this discrepancy might be that many FD patients visiting a general practitioner, dermatologist or cardiologist will not be diagnosed as FD.

Heterozygosity for E2 and the expression of FD

From the literature it is concluded that almost all FD patients exhibit the E2/E2 phenotype. Subjects with heterozygosity for the E*2 allele very rarely develop this disease. To our knowledge, only Rall et al. (1983) reported the existence of two FD patients (sibs) with the E3/E2 phenotype. We found three genetically unrelated FD patients with the E3/E2 phenotype among a total of 41 consecutive patients with FD. As FD patients with heterozygosity for the E*2 allele are very rare, whereas 15 percent of the general population exhibits heterozygosity for this allele, we wondered whether these three FD patients exhibit the common E2 (Arg 158 → Cys) isoform.

At present three different mutations have been described as giving rise to ApoE2: E2(Arg 158 → Cys), E2 (Arg 145 → Cys) and E2(Lys 146 → Gln). E2(Arg 158 → Cys is by far the most frequently occurring form of ApoE2. ApoE3 and E2(Lys 146 → Gln) contain one cysteine residue, whereas E2(Arg 158 → Cys) and E2(Arg 145 → Cys) contain two cysteine residues. This means that after cysteamine treatment, converting cysteine residues to a positively charged analogue of lysine, E3, E2(Arg 158 → Cys) and E2(Arg 145 → Cys) will focus at the E4 position, whereas the E2(Lys 146 → Gln) variant will focus at the E3 position (Weisgraber et al., 1982).

Fig. 2 represents an apoE immunoblot of an isoelectric focusing slab gel (pH range 5 to 7) applied with delipidated plasma samples from a normolipidemic E3/E2 heterozygote and a normolipidemic E3/E3 homozygote, as indicated. After cysteamine treatment of their plasma both apoE isoelectric focusing patterns changed into E4/E4. This indicates that in these plasma samples E2 contains two cysteine residues, whereas E3 contains one cysteine residue, as expected.

The same technique of cysteamine treatment was used for our FD patients. Fig. 2 shows the effect of cysteamine treatment on the ApoE isoelectric focusing pattern of one patient with the E2/E2 phenotype and one patient with the uncommon E3/E2 phenotype (as far as the FD is concerned). After cysteamine treatment, the E2/E2 pattern changed into E4/E4, indicating that also in this FD patient E2 contains two cysteine residues. Strikingly, after cysteamine treatment the ApoE isoelectric focusing pattern of the FD patient with the E3/E2 phenotype changed into E4/E3 instead of the E4/E4 pattern.

The results presented in Table 2 show that all three genetically unrelated familial dysbetalipoproteinemic patients (propositi) with the E3/E2 phenotype displayed the unexpected E4/E3 pattern after cysteamine treatment of their plasma. The cysteamine treatment technique was also used for 21 FD patients with the E2/E2 phenotype and for 50 apparently healthy subjects with the E3/E2 phenotype. All these subjects displayed the E4/E4 phenotype after cysteamine treatment of their plasma (Table 2).

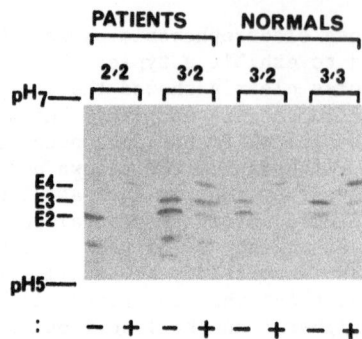

Fig. 2. ApoE phenotyping using the immunoblotting technique. Plasmas from two normolipidemic and two FD subjects were applied as indicated. Each plasma sample was treated with (+) and without (−) cysteamine.

Table 2. Apolipoprotein E phenotyping in E2/E2 and E3/E2 FD patients and in E3/E2 controls without (−) and with (+) cysteamine treatment.

Subjects	number analysed	Apo E phenotype	
		without (−)	with (+)
FD	21	2/2	4/4
	3	3/2	4/3
controls	50	3/2	4/4

It is known that degradation products of ApoE may focus in the E3 position. This problem is particularly relevant when the ApoE phenotyping is performed with the present immunoblotting technique using total plasma (Menzel & Utermann, 1986). Therefore, for the three patients with the uncommon E3/E2 phenotype the apoE phenotyping was also carried out by the conventional method using delipidated VLDL. From the results shown in Fig. 3 it is obvious that these three patients also exhibited the E3/E2 phenotype when isolated VLDL was used. Consequently, in these patients the presence of ApoE3 as determined by the immunoblotting method is not due to the presence of an ApoE degradation product in their plasma.

Our results indicate that the three genetically unrelated FD patients with the E3/E2 phenotype either contain an uncommon E2 isoform with only one cysteine residue or an uncommon E3 isoform that contains no cysteine residues. To discriminate between these two possibilities, we studied family members of these patients. In all three families the subjects with the E3/E3 phenotype displayed the E4/E4 phenotype after cysteamine treatment. This indicates that in the patients the E3 isoform normally contains one cysteine residue. Consequently, they exhibit an uncommon E2 isoform that contains only one cysteine residue.

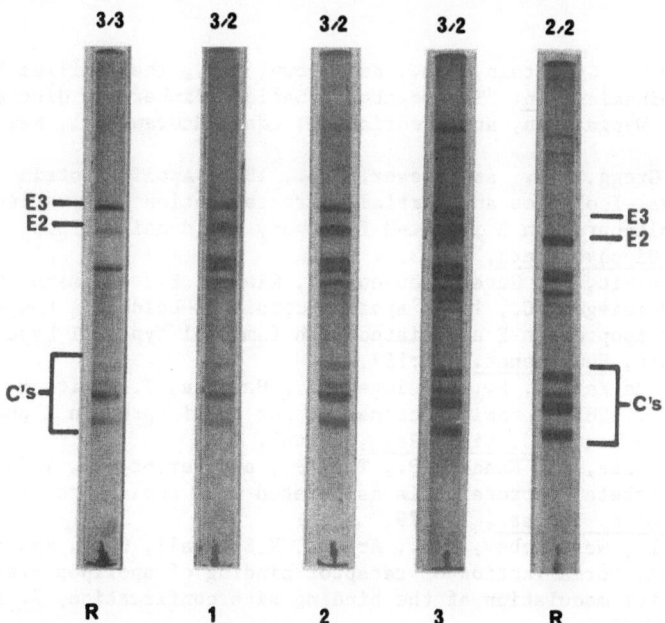

Fig. 3. Apolipoprotein E phenotyping with the conventional method using isolated VLDL. Numbers 1, 2 and 3 represent the three FD patients with the E3/E2 phenotype as determined by the immunoblotting technique. R stands for reference samples, i.e. subject with E3/E3 homozygosity (left) and E2/E2 homozygosity (right).

In addition, family studies showed that in these families the uncommon E2 allele cosegrates with FD in the first and second generation. In the third generation of these families the subjects are still too young to express FD, which usually develops after the third decade.

At present, it is unknown whether these patients with the E3/E2 phenotype exhibit the same uncommon E2 allele. Two rare E2 alleles have been described so far, E2(Arg 145 → Cys) and E2(Lys 146 → Gln) (Rall et al., 1982; Rall et al., 1983). The presence in these E3/E2 patients of the E*2(Arg 145 → Cys) allele can be excluded, as the E2(Arg 145 → Cys) variant contains two cysteine residues and would therefore focus at the E4 position after cysteamine treatment. Rall et al. (1983) reported the existence of two FD patients (sibs) with the E3/E2(Lys 146 → Gln) phenotype. Whether the E2(Lys 146 → Gln) variant that contains one cysteine residue is also present in our E3/E2 patients, is currently under investigation.

Because of the fact that FD patients with the E3/E2 phenotype are very rare and the fact that our three E3/E2 patients all exhibit an uncommon E2 allele, instead of the by far most frequently occurring E2(Arg 158 → Cys) allele, we would like to suggest that this uncommon E*2 allele behaves like a dominant trait in the expression of familial dysbetalipoproteinemia.

ACKNOWLEDGEMENTS

These investigations were supported by the Division for Health Research TNO (project no. 900-504-059).

REFERENCES

Fredrickson, D.S., Goldstein, J.L., and Brown, M.S., The familial hyper-lipoproteinemias, in: "The metabolic basis of inherited disease," Stanbury, Wyngaarden, and Fredrickson, eds., McGraw-Hill, New York (1980).

Ghiselli, G., Gregg, R.E., and Brewer, H.B., 1984, Apolipoprotein E-Bethesda. Isolation and partial characterization of a variant of human apolipoprotein E isolated from very low density lipoproteins, Biochim. Biophys. Acta, 794:333.

Havekes, L., de Wit, E., Gevers Leuven, J., Klasen, E., Utermann, G., Weber, W., and Beisiegel, U., 1986, Apolipoprotein E3-Leiden. A new variant of human apolipoprotein E associated with familial type III hyperlipo-proteinemia, Hum. Genet., 73:157.

Havekes, L.M., de Knijff, P., Beisiegel, U., Havinga, J., Smit, M., and Klasen, E., 1987, A rapid micromethod for apolipoprotein E phenotyping directly in serum, J, Lipid Res., 28:455.

Havel, R.J., Kotite, L., Kane, J.P., Tun, P., and Bersot, T., 1983, A typical dysbetalipoproteinemia associated with apolipoprotein phenotype E3/3, J. Clin. Invest., 72:379.

Innerarity, T.L., Weisgraber, K.H., Arnold, K.S., Rall, S.C., and Mahley, R.W., 1984, Normalization of receptor binding of apolipoprotein E2. Evidence for modulation of the binding site confirmation, J. Biol. Chem., 259:7261.

Klasen, E.C., Smit, M., de Knijff, P., Gevers Leuven, J., Kempen-Voogd, R., and Havekes, L., 1987, Apolipoprotein E phenotype and gene distribution in The Netherlands, Hum. Hered., 37:340.

Menzel, H.J., and Utermann, G., 1986, Apolipoprotein E phenotyping from serum by Western blotting, Electrophoresis, 7:492.

Rall, S.C., Weisgraber, K.H., and Mahley, R.W., 1982, Human apolipoprotein E. The complete amino acid sequence, J. Biol. Chem., 257:4171.

Rall, S.C., Weisgraber, K.H., Innerarity, T.L., Bersot, T.P., and Mahley, R.W., 1983, Identification of a new structural variant of human apo-lipoprotein E, E2(Lys 146 → Gln), in a type III hyperlipoproteinemic subject with the E3/E2 phenotype, J. Clin. Invest., 72:1288.

Schneider, W.J., Kovanen, P.T., Brown, M.S., Goldstein, J.L., Utermann, G., Weber, W., Havel, R.J., Kotite, L., Kane, J.P., Innerarity, T.L., and Mahley, R.W., 1981, Familial dysbetalipoproteinemia. Abnormal binding of mutant apoprotein E to low density lipoprotein receptors of human fibroblasts and membranes from liver and adrenal of rats, rabbits, and cows, J. Clin. Invest., 68:1075.

Sherrill, B.C., Innerarity, T.L., and Mahley, R.W., 1980, Rapid hepatic clearance of the canine lipoproteins containing only the E apoprotein by a high affinity receptor, J. Biol. Chem., 255:1804.

Utermann, G., Steinmetz, A., and Weber, W., 1982, Genetic control of human apolipoprotein E polymorphism: comparison of one- and two-dimensional techniques of isoprotein analysis, Hum. Genet., 60:344.

Weisgraber, K.H., Innerarity, T.L., and Mahley, R.W., 1982, Abnormal lipo-protein receptor binding activity of the human E apoprotein due to cysteine-arginine interchange at a single site, J. Biol. Chem., 257:2518.

Yamamura, T., Yamamoto, A., Hiramori, K., and Nambu, S., 1984, A new isoform of apolipoprotein E - apo E5 - associated with hyperlipidemia and atherosclerosis, Atherosclerosis, 50:159.

Zannis, V.I., and Breslow, J.L., 1981, Human very low density lipoprotein apolipoprotein E isoprotein polymorphism is explained by genetic variation and posttranslational modification, Biochemistry, 20:1033.

IDENTIFICATION OF AN ITALIAN KINDRED WITH A VARIANT APOLIPOPROTEIN E (E1) ASSOCIATED WITH TYPE III HYPERLIPOPROTEINEMIA

C. Gabelli, G. Baggio, *A. Pagnan, *G. Zanetti, G.M. Barbato, S. Martini, C. Bilato, C. Corti and G. Crepaldi

Departments of Internal Medicine and *Clinical Medicine, University of Padova, Italy

INTRODUCTION

Human apolipoprotein E (apo E) is a 299 amino acid polypeptide associated with circulating lipoproteins. During the last decade apo E has been extensively investigated. The complete amino acid sequence has been determined[1] and more recently also the gene organization[2,3,4], cDNA and mRNA sequences have been described[5], showing important similarities with other apolipoproteins.

Utermann[6,7] first showed a microheterogeneity of apo E after isoelectric focusing due to a polymorphism of the protein. Three common apo E isoforms may be isolated from plasma: E2, E3 and E4. They differ from each other by one charge unit due to an arginine-cysteine interchange at position 112 and 158[8]. The three isoproteins are under the control of three alleles, designated $\varepsilon 2$, $\varepsilon 3$ and $\varepsilon 4$, inherited in codominant fashion in one autosomal locus[8]. Six different genotypes and phenotypes are therefore generally found: three homozygous, E2/2, E3/3, E4/4 and three heterozygous, E2/4, E3/4, E2/3.

Apo E is thought to mediate the rapid clearance of triglyceride-rich lipoprotein remnants[9], in addition it binds to the high affinity LDL receptor[10] and a modulating role of apo E in the low density lipoprotein (LDL) metabolism is possible[11,12]. An abnormal function of apo E, as in apo E2 homozygousity[13] and in rare apo E variants[14-19], may determine an impaired lipoprotein metabolism, which often leads to a type III hyperlipoproteinemia (HLP). This disease is characterized by hypertriglyceridemia, hypercholesterolemia, decreased catabolism of chylomicron remnants which accumulate in plasma as beta migrating VLDL (very low density lipoproteins). The patients with type III HLP usually have decreased levels of LDL, often palmar xanthomas and premature atherosclerosis[20].

In 1984 Weisgraber et al. described a Finnish kindred carrying a variant apo E (E1) associated with severe hypertriglyceridemia[21]; we describe here an Italian family with the presence of a similar apo E isoform, but associated with type III HLP and severe atherosclerosis. at a concentration of 1 mg of cysteamine/100 µg total VLDL protein.

METHODS

Patients

Eighteen subjects of the family (Fig. 1) participated in the study; all of them underwent physical examination, lipoprotein and apo E phenotyping. Selected subjects underwent a study of the cardio-vascular condition by non-invasive methodologies (ECG, Doppler ultra-sonography).

Clinical evaluation

The proband for this apo E variant was discovered during the determination of the apo E phenotypes of a large group of hyperlipidemic subjects. Apo E and the lipoproteins were evaluated after discontinuation of lipid-lowering drugs and unresctricted diet for 4 weeks. Clinical evaluation of the proband was performed during hospitalization in the ward of the Department of Internal Medicine of the University of Padova.

Lipid and Lipoprotein separation

Plasma was obtained from each subject after an overnight fast in EDTA and the lipoproteins were isolated by sequential ultracentri-fugation using KBr to adjust the density[22]. Triglycerides and total cholesterol were determined by enzymatic procedures[23,24]. Apo E was quantitated by radial immunodiffusion[25] using commercial plates (Daiichi Pure Chemicals Co., Tokio, Japan).

Apo E and other apoprotein characterization

Apo E phenotypes were obtained by analytical isoelectric focusing using a modification of the original method of Warnick[27,28]. VLDL (d < 1.006 g/ml) containing 40–150 μg of total protein[29] were desalted by chromatography on PD-10 columns containing Sephadex G-25 (Pharmacia, Uppsala, Sweden) equilibrated with 10 mM ammonium bicarbonate, 0.01% Na_2 EDTA, 2 mM NaN_3, pH 8.2. The samples were then liophylized and delipidated with chloroform/methanol (2:1 v/v) followed by an additional extraction with chloroform/methanol/ethyl ether (3:1:3 v/v). The protein pellet was dried under nitrogen stream and dissolved in 8 M Urea containing 0.1 M DTT (Fluka, RFT) and 3:100 (v/v) Pharmalite 4–6.5 (Pharmacia). Protein separation was carried out in vertical slab poly-acrylamide gels (7.5% T, 2.6% C), 1.5 mm thick, pH range 4–6.5 after focusing for 4 hrs with 15 watt/gel using a Protean Bio-Rad apparatus. Protein bands were stained in a solution of 0.025% Coomassie G-250 (Sigma) in 3.5% perchloric acid. Two-dimensional electrophoresis was performed on VLDL as previously described[30]. For neuraminidase treatment of apo E 100 μg of VLDL protein were incubated in 0.1 M potassium monophate, 1 mM calcium chloride, KOH pH 5.0, incubated for 4 hrs at 37°C with 1 U neuraminidase type X (Sigma). Cysteine modification of apo E by cysteamine was performed on total VLDL in 10 mM ammonium bicarbonate incubated for 12 hrs at room temperature. Cysteamine (Sigma)

182

RESULTS

The proband R.E. is a 61 year old female, who was referred to our Lipid Clinic in 1987 for a persistent hypercholesterolemia and hypertriglyceridemia. Her clinical history revealed that at the age of 45 she noticed the first palpebral xanthomas. Physical examination showed an overweight (ht 170 cm, wt 100 Kg), multiple palpebral xanthomas and a tendon xanthoma on the extensor of one finger. She had a mild hypertension and a mild type II diabetes without ocular or kidney complications. A non-invasive evaluation of the vascular tree revealed multiple atherosclerotic plaques of the internal and external carothid

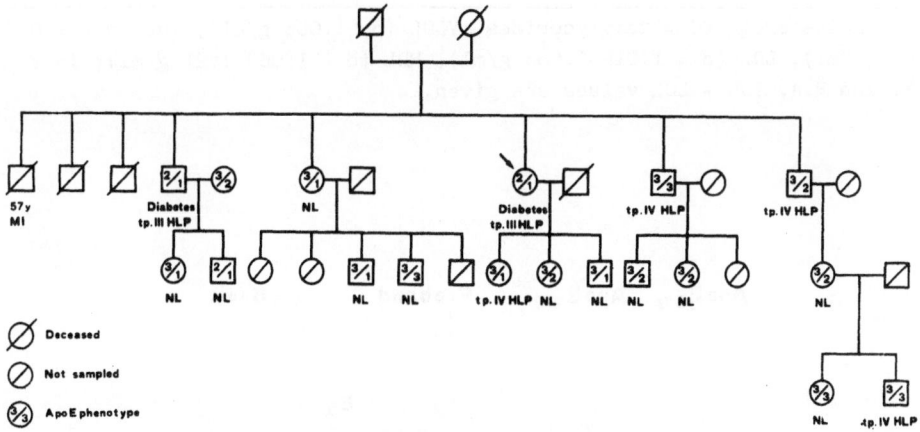

Fig. 1. Complete genealogic tree of the family. The arrow indicates the proband. NL stands for normolipidemic subject while carriers of hyperlipidemia (HLP) are indicated. A subject who died at the age of 57 for a miocardial infarction (MI) is shown.

arteries, multiple lesions of the aortha, iliac and femoral arteries with occlusion of the left superficial femoral artery. Although she had no symptoms of cardiovascular disease, the ECG was suggestive of ischemic heart disease. The lipoprotein values of the patient and of her offspring are given in table 1. She exhibited type III hyperlipoproteinemia as established by the lipoprotein pattern, a VLDL cholesterol/plasma triglycerides ratio greater than 0.3 and the presence of a broad beta-band by lipoprotein electrophoresis. The plasma apo E level was increased (22 mg/dl). Apo E analysis by isoelectric focusing revealed two bands in position E2 and E1 (Fig. 2). This finding was confirmed by two-dimensional electrophoresis and immunoblot (data not shown). The treatment of VLDL with neuraminidase did not change

Table 1. Lipid and lipoprotein values and apo E phenotype of the proband, of her offspring and of one normolipidemic subject with apo E2/1 phenotype.

	Sex	Age (yrs)	Plasma		VLDL	IDL	LDL	HDL	ApoE phen.	VLDL·CH/ Plasma·TG ratio
			CH	TG	CH	CH	CH	CH		
			mg/dl							
Proband (R.A.)	F	61	448	641	265	60	50	36	2/1	0.41
V.A.	M	35	170	155	22	6	102	35	3/1	0.14
V.M.	F	30	182	85	7	3	110	35	3/2	0.08
V.A.M.	F	39	209	306	49	– 110 –		42	3/1	0.16
R.A.	M	24	113	68	10	– 55 –		43	2/1	0.14

CH = cholesterol, TG = triglycerides, VLDL (d< 1.006 g/ml), IDL (d = 1.006–1.019 g/ml), LDL (d = 1.019–1.063 g/ml), HDL (d = 1.063–1.21 g/ml); in V. A.M. and R.A. IDL + LDL values are given.

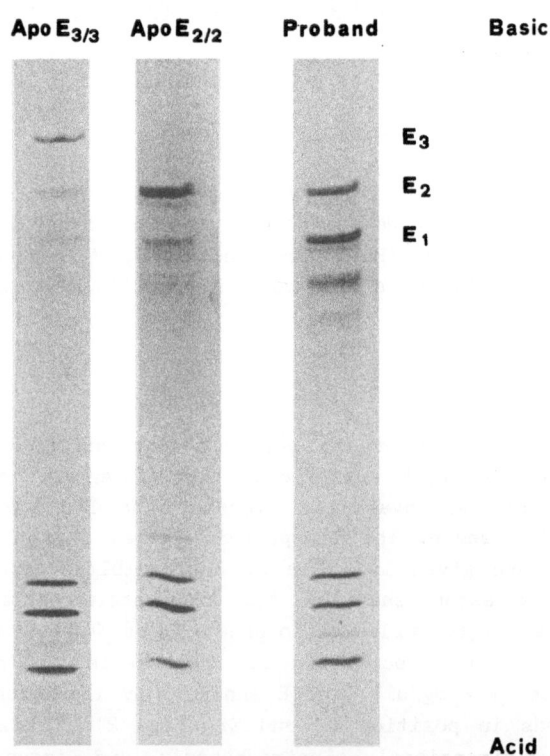

Fig. 2. Isoelectric focusing of the proband apo VLDL compared with a subject with apo E2/2 phenotype and a subject with apo E3/3 phenotype.

the apo E phenotype and ruled out the hypothesis that the apo E1 band might be related with an abnormal glycosilation of the apolipoprotein. To further understand the nature of the variant isoprotein, VLDL obtained from the proband and other members of the family were incubated with cysteamine (Fig. 3). The chemically modified isoproteins E2 and E3 both migrated in position E4 because of the addition by cysteamine of one positive charge for every cysteine residue present on the molecule. The variant isoprotein E1 shifted two units of charge (position E3) by cysteamine treatment (Fig. 3). This observation indicates that the molecule of apo E1 contains two cysteine residues. Analysis of other family members showed that a brother died at the age of 57 by myocardial infarction. Another brother presented diabetes, type III HLP associated with E2/1 phenotype and severe periferal vascular disease. A third young subject with apo E2/1 phenotype was normolipidemic and five other relatives were apo E3/1 etherozygotes. No E1 homozygote was found. Four family members showed type IV HLP, but no clear association with the presence of apo E1 was observed.

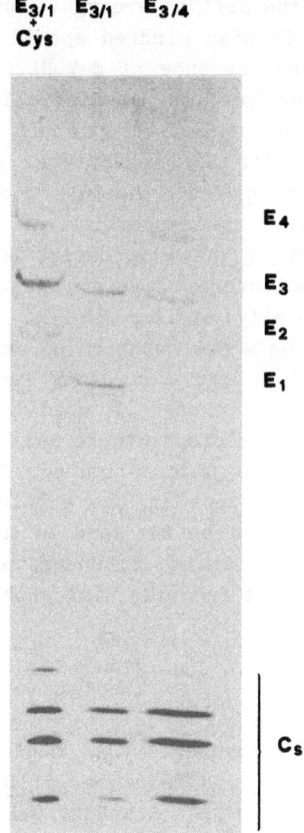

Fig. 3. Chemical modification of apo E. VLDL obtained from a subject with apo E3/1 phenotype (central lane) were treated with cysteamine (left). Apo E1 shows a shift to position E3 after the treatment. A subject with an apo E3/4 phenotype is shown (right) for comparison.

DISCUSSION

A number of new apolipoprotein E variants have been described in the recent years. These observations are important mainly because they can clarify the relationship between the primary structure and function of the apoprotein. However, it should also be remarked that some of these new mutants might not be unique and, as rare alleles spread throughout the population, they might contribute to the genetic variability of lipoprotein levels.

In this study we described an Italian kindred with an unusual isoprotein E with E1 mobility. Neuraminidase treatment ruled out the possibility that an abnormal glycosilation could cause the presence of the band whose position is usually occupied by sialilated forms of E3 and E2. Moreover, after chemical modification, it appeared that the variant E1 contained two cysteine residues. Similar properties had been previously found in two other apo E variants with E1 mobility: apo E Bethesda[16] and apo E1 described by Weisgraber[21] et al. However, it should be pointed out that the clinical characteristics in these two kindreds were different if compared with the Padova kindred. In the Bethesda E1 both the E2/1 heterozygote and an E3/1 subject presented type III HLP. In the Finnish kindred apo E1 was associated with hyper-triglyceridemia and the presence of ß-VLDL. Both the described families had few members, therefore it is difficult to fully understand the clinical impact of the variant isoprotein. In the Padova kindred we found three subjects with E2/1 phenotype. Only two of them presented type III HLP, indicating that the age and the presence of diabetes might be important for the expression of the disease. Most subjects with E3/1 phenotype were normolipidemics and some of them presented low total and LDL cholesterol levels. These features make apo E1, in this family, functionally similar to the usual E2 and suggest that apo E1 might derive from a new mutation of an E2 allele.

Recent reports[30,31] described other families with apo E1 and type III HLP. The methodology presently used in the determination of apo E phenotype may not always discriminate well between apo E1 and apo A-I. Apo E1 could therefore be underestimated. It is possible that the ε1 allele is relatively frequent in the population and we speculate that the allele for apo E1 might be the same in the majority of the described kindreds. Further studies aiming to understand the molecular alteration of apo E1 in the different families will probably answer this question.

REFERENCES

1. S.C. Rall, K.H. Weisgraber, R.W. Mahley, human apolipoprotein E: the complete amino acid sequence, J. Biol. Chem. 257:4171 (1982).
2. Y.K. Paik, D.J. Chang, C.A. Reardon, G.E. Davies, R.W. Mahley, J.M. Taylor, Nucleotide sequence and structure of the human apolipo-protein E gene, Proc. Natl. Acad. Sci. USA 82:3445 (1985).
3. A.K. Das, J. McPherson, G.A. Bruns, S.K. Karathanasis, J.L. Breslow, Isolation, characterization and mapping to chromosome 19 of the human apolipoprotein E gene, J. Biol. Chem. 260:6240 (1985).

4. J. Scott, T.J. Knott, D.J. Shaw, J.D. Brook, Localization of genes encoding apolipoprotein CI, CII and E to the p13--cen region of human chromosome 19, Hum. Genet. 71:44 (1985).

5. J.W. McLean, N.A. Elshourbagy, D.J. Chang, R.W. Mahley, J.M. Taylor, Human apolipoprotein E mRNA, cDNA cloning and nucleotide sequencing of a new variant, J. Biol. Chem. 259:6498 (1984).

6. G. Utermann, Isolation and partial characterization of an arginine-rich apolipoprotein from human plasma very-low-density lipoproteins: apolipoprotein E, Hoppe-Seyler's Z. Physiol. Chem. 356:1113 (1975).

7. G. Utermann, M. Hees, A. Steinmetz, Polymorphism of apolipoprotein E and dysbetalipoproteinemia in man, Nature 269:604 (1977).

8. K.H. Weisgraber, S.C. Rall Jr., R.W. Mahley, Human E apoprotein heterogeneity. Cysteine-arginine interchange in the amino acid sequence of the apoE isoforms, J. Biol. Chem. 256:9077 (1981).

9. B.C. Sherrill, T.L. Innerarity, R.W. Mahley, Rapid hepatic clearance of the canine lipoproteins containing only the E apoprotein by a high affinity receptor, J. Biol. Chem. 255:1804 (1980).

10. R.W. Mahley, D.Y. Hui, T.L. Innerarity, K.H. Weisgraber, Two independent lipoprotein receptors on hepatic membranes of dog, swine and man, J. Clin. Invest. 68:1197 (1981).

11. C. Gabelli, R.E. Gregg, L.A. Zech, E. Manzato, H.B. Brewer Jr., Abnormal low density lipoprotein metabolism in apolipoprotein E deficiency, J. Lipid Res. 27:326 (1986).

12. R.E. Gregg, L.A. Zech, C. Gabelli, J.M. Hoeg, H.B. Brewer Jr., The role of apolipoprotein E and the low-density lipoprotein receptor in modulating the in vivo metabolism of apolipoprotein-B-containing lipoproteins, in: "Cardiovascular Disease: molecular and cellular mechanism, prevention and treatment," L. Gallo, ed., Plenum, New York (1987).

13. W.J. Schneider, P.T. Kovanen, M.S. Brown, J.L. Goldstein, G. Utermann, W. Weber, R.J. Havel, L. Kotite, J.P. Kane, T.L. Innerarity, R.W. Mahley, Familial Dysbetalipoproteinemia. Abnormal binding of mutant apoprotein E to low density lipoprotein receptors of human fibroblasts and membranes from liver and adrenals of rats, rabbits, and cows, J. Clin. Invest. 68:1075 (1981).

14. R.J. Havel, L. Kotite, J.P. Kane, P. Tun, T.P. Bersot, Atypical familial dysbetalipoproteinemia associated with apolipoprotein phenotype E3/3, J. Clin. Invest. 72:379 (1983).

15. S.C. Rall, K.H. Weisgraber, T.L. Innerarity, T.P. Bersot, R.W. Mahley, C.B. Blum, Identification of a new structural variant of human apolipoprotein E, E2 (Lys$_{146}$ $\rightarrow$ Glu) in a type III hyper-lipoproteinemic subject with the E3/2 phenotype, J. Clin. Invest. 72:1288 (1983).

16. R.E. Gregg, G. Ghiselli, H.B. Brewer Jr., Apolipoprotein E Bethesda: a new variant apolipoprotein E associated with type III hyper-lipoproteinemia, J. Clin. Endocrinal Metab., 57:969 (1983).

17. L. Havekes, E. de Wit, J. Gevers Leuven, E. Klasen, G. Utermann, W. Weber, U. Beisiegel, Apolipoprotein E3-Leiden. A new variant of human apolipoprotein E associated with familial type III hyper-lipoproteinemia, Hum. Genet., 73:157 (1986).

18. M.R. Wardell, S.O. Brennan, E.D. Janus, R. Fraser, R.W. Carrell, Apolipoprotein E2-Christchurch (136 Arg $\rightarrow$ Ser). New variant of

human apolipoprotein E in a patient with type III hyperlipo-proteinemia, J. Clin. Invest., 80:483 (1987).

19. E.J. Schaefer, R.E. Gregg, G. Ghiselli, T.M. Forte, J.M. Ordovas, L.A. Zech, H.B. Brewer Jr., Familial apolipoprotein E deficiency, J. Clin. Invest. 78:1206 (1986).

20. H.B. Brewer Jr., L.A. Zech, R.E. Gregg, D. Schwartz, E.J. Schaefer, Type III hyperlipoproteinemia: diagnosis, molecular defect, pathology and treatment, Ann. Intern. Med. 98:623 (1983).

21. K.H. Weisgraber, S.C. Rall, T.L. Innerarity, R.W. Mahley, T. Kuusi, C. Ehnholm, A novel electrophoretic variant of human apolipo-protein E. Identification and characterization of apolipoprotein E1, J. Clin. Invest., 73:1024 (1984).

22. R.J. Havel, H.A. Eder, J.H. Bragdon, The distribution and chemical composition of ultracentrifugally separated lipoproteins in human serum, J. Clin. Invest. 34:1345 (1955).

23. A.W. Wahlefeld, Triglycerides determination after enzymatic hydrolysis, in: "Methods of Enzymatic Analysis," H.U. Bergmeyer, ed., Academic Press Inc., New York (1976).

24. C. Allen, S. Poon, C. Chan, W. Richmond, P. Fu, Enzymatic determi-nation of total cholestrol, Clin. Chem. 20:470 (1974).

25. G. Mancini, A.O. Carbonara, J.F. Heremans, Immunochemical quanti-tation of antigens by single radial immunodiffusion, Immunochemistry 2:235 (1965).

26. G.R. Warnick, C. Mayfield, J.J. Albers, W.R. Azzard, Gel isoelectric focusing method for specific diagnosis of familial hyperlipo-proteinemia type III, Clin. Chem. 25:279 (1979).

27. C. Gabelli, G. Baggio, S. Martini, G.M. Barbato, C. Bilato, S. Pigozzo, L. Previato, M.C. Corti, G. Crepaldi, A contribution to the genetic characterization of primary hyperlipoproteinemias: the separation of the apolipoprotein E isoforms by isoelectric focusing, Prog. Med. Lab. 2:261 (1988).

28. O.H. Lowry, N.J. Rosenbrough, A.L. Farr, R.J. Randall, Protein measurement with Folin phenol reagent, J. Biol. Chem. 193:265 (1951).

29. D.L. Sprecher, L. Tamm, H.B. Brewer Jr., Two-dimensional electro-phoresis of human plasma apolipoproteins, Clin. Chem. 30:2084 (1984).

30. D. Seidel, H. Wieland, C. Ruppert, Improved techniques for assessment of plasma lipoprotein patterns. I. Precipitation in gels after electrophoresis with polyanionic compounds, Clin. Chem. 19:737 (1973).

31. J.M. Ordovas, L. Litwack-Klein, P.W.F. Wilson, M.M. Schaefer, E.J. Schaefer, Apolipoprotein E isoform phenotyping methodology and population frequency with identification of apo E1 and apo E5 isoforms, J. Lipid Res. 28:371 (1987).

32. A. Steinmetz, N. Assefbarkhi, C. Eltze, H. Funke, G. Assman, H. Kaffarnik, Homozygousity for a rare apolipoprotein E (Apo E-1) mutant in a patient with type III hyperlipoproteinemia and a normolipemic subject with dysbetalipoproteinemia, Eur. J. Clin. Invest. 18-II:A12 (1988).

NORMOLIPIDEMIC DYSLIPOPROTEINEMIA IN PATIENTS WITH CORONARY ARTERY DISEASE

Giancarlo Ghiselli, Ellison Wittels, Jacques
Heibig, and Antonio M. Gotto jr.

Baylor College of Medicine and The Methodist
Hospital, Department of Medicine and The Veteran
Administration Hospital, Houston, TX

INTRODUCTION

Evidence has been accumulated in the past years that there
is a significant relationship between the concentration of
circulating plasma cholesterol and the development of
atherosclerosis. The results of three large prospective
studies namely, the Framingham Heart Study (1), the Pooling
Project (2), and the Multiple Risk Factor Intervention Trial
(3), show that the relationship between plasma cholesterol
and Coronary Artery Disease (CAD) is continous and the risk
of premature atherosclerosis rise dramatically with plasma
cholesterol concentration above 250 mg/dl. Furthermore the
recently concluded Lipid Research Clinic Primary
Intervention Trial (4), The Helsinki Heart Study (5) and the
Cholesterol-Lowering Atherosclerosis Study (6), all show
benefit in terms of mortality by myocardial infarction
resulting from reduction with hypolipidemic agents of the
plasma concentration of cholesterol.

In humans, the majority of the plasma cholesterol is carried
by the low density lipoproteins (LDL) (7). The work of
Goldstein and Brown and others on the receptor-mediated
regulation of LDL plasma concentration, has lead to the
elucidation of the ethiology of most forms of
hypercholesterolemia and of the causal relationship existing
between elevated plasma cholesterol levels and increased
risk of atherosclerosis. In the model of LDL interaction
with its receptor, the protein moiety of LDL, namely
apolipoprotein B-100 (apoB-100) is the molecular determinant
allowing recognition and binding of LDL at the cell surface.
LDL receptor occupancy, is followed by cell internalization
and degradation. In this way apoB-100 is a major
determinant for LDL and cholesterol level in plasma (8).
Not surprisingly a number of epidemiological studies show

TABLE I

Plasma and Lipoprotein Lipid Concentration (mg/dl)
in Subjects with CAD and in Age Matched Controls

Subjects	Total Chol	Total TG	VLDL°		LDL		HDL	
			Chol	TG	Chol	TG	Chol	TG
Controls (12)	185 ± 8*	80 ± 10	7 ± 1	28 ± 7	126 ± 4	40 ± 3	46 ± 3	22 ± 2
CAD (17)	192 ± 8	97 ± 8	11 ± 1**	44 ± 4**	121 ± 3	45 ± 1	38 ± 4**	22 ± 1

o Lipids were quantitated on VLDL (d < 1.006 g/ml) LDL (d = 1.006–1.006 g/ml)
and HDL (d = 1.063–1.210 g/m) isolated by ultracentrifugation

* Mean ± SD

** p < 0.05 vs Controls

there is a strong positive correlation between the level in
plasma of (apoB-100) and the incidence of CAD (9). In fact
this correlation, may be better than that based on plasma
cholesterol or LDL cholesterol concentrations alone (10).
Sniderman and collegues (11), have described a disease
named hyperapobetalipoproteinemia strongly associated to
premature CAD which is characterized by an increased number
of circulating LDL particles with increased apoB-100 to
cholesterol ratio. Whether or not these and other abnormal
lipoprotein particles are causative agents for
atherosclerosis even when plasma cholesterol level is
within the normal range, is not known; for example there are
no data at present showing a direct correlation between
abnormal LDL particle concentration and risk for CAD as it
has been demonstrated for plasma cholesterol. Beyond these
uncertainties, the fact remains that the vast majority of
the subjects with CAD have plasma lipid levels well within
the normal limits and the idea that in these subjects LDL
composition and metabolism may be abnormal causing premature
atherosclerosis, has prompted a number of studies.

METABOLISM OF APOB-100 IN NORMAL AND DYSLIPOPROTEINEMIC
STATES

ApoB-100 is a glycoprotein with a polypeptide chain of 4,536
aminoacids (12). It is the form of apoB synthesized by the
liver (13). The intestine may also secrete apoB-100 but
apoB-48 is the major form of apoB synthesized in this organ
(14). ApoB-100 is the major structural protein of both
VLDL, IDL and LDL. It reaches circulation mostly as a VLDL
component but some IDL and LDL are also directly secreted by
the liver (15). Each newly secreted VLDL particles contains
one molecules of apoB-100 that is retained in VLDL remnants
(IDL) and the final plasma delipidization product LDL.
Radiolabelled apoB-100 is then suited for monitoring the
metabolic interrelationships existing between these three
lipoprotein classes. It is not definetively known at the
present whether apoB-100 in VLDL remains the same in IDL and
LDL. It is assumed so based on the fact that on SDS-PAG
electrophoresis, apoB-100 of VLDL, IDL and LDL has the same
apparent molecular weight (14). Moreover direct aminoacid
sequencing of fragments of apoB-100 isolated from LDL,
matches that deduced from nucleotide sequencing of liver
apoB-100 cDNA (16). On the other hand, there are mounting
evidences that apoB-100 may undergo covalent modification in
vivo. These changes include oxidation (17) and
malondialdehyde modification (18). In addition the sugar
moiety of apoB-100 in VLDL, IDL and LDL has yet to be
carefully characterized.

There is now overwhelming evidence that apoB-100 metabolism
is altered in a number of hyperlipoproteinemic conditions

TABLE II

VLDL and LDL ApoB-100 and Plasma ApoA-I and ApoA-II Concentration
(mg/dl) in Subjects with CAD and in Age Matched Control

Subject	VLDL-apoB-100	LDL-apoB-100	ApoA-I	ApoA-II
Controls (12)	6 ± 1*	67 ± 7	100 ± 4	30 ± 1
CAD (47)	10 ± 1**	74 ± 4	83 ± 2**	26 ± 1**

*   Mean ± SD

** p < 0.05 vs Control

(reviewed by Grundy and collegues in ref. 19 and 20). In familial hypercholesterolemia LDL apoB-100 production rate is increased in part due to direct LDL secretion by the liver. Furthermore LDL fractional catabolic rate is significantly decreased. On the other hand, the production rate and the fractional catabolic rate of VLDL apoB-100 are normal. These data support the concept that elevated concentration of LDL in plasma of familial hypercholesterolemic patients are secondary to LDL oversynthesis and to a defect in the receptor responsible for the catabolism of LDL at the tissues. In other patients, hypercholesterolemia may be caused by overproduction of VLDL apoB-100, to structural defect(s) of the LDL receptor leading to lower affinity for the ligand, or to structural mutation of apoB-100. VLDL overproduction is common in patients with hypertriglyceridemia. In this case VLDL apoB-100 overproduction is concomitant to either overproduction or defective clearance of the triglycerides. Type IIb, IV, and V hypertriglyceridemias are frequently the resulting phenotype. In addition the VLDL of these subjects are poorly catabolized and a large fraction of VLDL leaves the plasma space by direct tissue catabolism. LDL lipid composition is also abnormal in patients with hypertriglyceridemia, and this may contribute to their accelerated catabolism (21). In subjects with type III hyperlipoproteinemia the effect of elevated VLDL synthesis is exacerbated by a functionally abnormal apoE mutant (apoE-2) (22). ApoE is important in the plasma clearance of chylomicron and VLDL remnants through the liver B/E receptor. Since apoE-2 interacts poorly with this receptor, remnants lipoproteins accumulates (22).

Overproduction of VLDL and/or other apoB-100 containing lipoproteins not only affects apoB-100 concentration and distribution in plasma but also has important consequences on the metabolism of HDL. In hypertriglyceridemic patients in particular, there is an inverse relationship between the catabolic rate of HDL and that of VLDL (24,25). This correlation has been explained in terms of perturbances that prolonged residence of VLDL in plasma may have on HDL metabolism (26). Evidences to date suggest that sluggish removal of VLDL in hypertriglyceridemic patients, leads to increased triglycerides mass transfer from these lipoproteins to LDL and HDL which become enriched with triglyceride. This process of lipid transfer is thought to be mediated by a lipid transfer protein (27). Overeactivity of the lipid transfer protein reaction - as it has been proposed by Eisenberg (28) - is thus responsible for the abnormal composition of the lipoproteins in hypertriglyceridemic patients. In HDL this causes increased apoA-I and apoA-II catabolism lowering total HDL plasma concentration in particular that of the HDL-2 subpopulation (26,28). Hypoalfalipoproteinemia and HDL subpopulation derangements are also well documented in patients with CAD

TABLE III

Lipid and Apolipoprotein Concentration (mg/dl) in
the Subjects Partecipating to the Turnover Studies

Patients	Total TG	Total C	HDL-C	ApoA-I	VLDL-apoB-100	LDL-apoB-100
Controls (47-67; n=3)§	140 ± 57*	199 ± 42	53 ± 10	105 ± 3	5 ± 2	56 ± 5
Young Volunteers (23-29; n=3)	98 ± 21	155 ± 18	54 ± 4	110 ± 1	4 ± 1	63 ± 11
CAD (48-69; n=5)	131 ± 28	195 ± 14	37 ± 3**	80 ± 4**	9 ± 2**	87 ± 7
HTG (54-72; n=4)	346 ± 64**	201 ± 22	22 ± 8**	77 ± 4**	16 ± 3**	61 ± 10

§   In parenthesis the age range and the number of subjects
*   Mean ± SD
** p < 0.05 vs Control

which are, in their majority, normolipidemic (29).
Conceivably also in these normolipidemic patients HDL
abnormalities herald defect in lipoprotein metabolism. In
spite of the importance of this subject as it may broaden
the knowledge on the pathogenesis of atherosclerosis and of
the role of plasma lipoproteins, interest in the past has
been scarse (30-32). We thus begun investigating the
metabolism of apoB-100, apoA-I and apoA-II in patients with
coronary atherosclerosis that were normolipidemic and had
low plasma apoA-I concentration. A cutpoint of 90 mg/dl was
selected as this value is the median of apoA-I concentration
in our group of CAD patients and correspond to the lower
quintile of the apoA-I values in an age matched group of
healthy individuals. In these subgroup of
dyslipoproteinemic patients we addressed the question of
which are the lipoprotein defects leading to decreased HDL
and possibly to premature CAD.

METABOLISM OF APOB-100 IN NORMOLIPIDEMIC SUBJECTS WITH CAD

Subjects for the turnover studies were recruited among a
larger group of patients that had undergone angiography and
found having or not clinically relevant coronary
atherosclerosis defined as more than 50% obstruction at
least one major coronary artery. In this population
constituted by male Caucasian in their 40's to 60's,
normolipidemic subjects without diabetes, other metabolic
disease or pulmonary or kidney diseases, where 76%. At the
time of the angiography these patients were not taking
hypolipidemic drugs. Most of the subjects were however
taking cardiovascular acting agents such as beta-blockers,
diuretics, calcium antagonists and nitrates either alone or
in combination. Beta-blockers and diuretics have some
effect on plasma lipids but the patients taking these agents
on the average did not had lipid and apolipoprotein plasma
levels significantly different from those of the patients
taking other drugs or none at all. Plasma and lipoprotein
cholesterol and triglyceride concentration are reported in
Table I. Subjects without CAD and those with CAD had
similar plasma cholesterol levels. Triglyceride
concentration was slightly elevated in the patients with
CAD. Quantitation of the lipid concentration of the
different lipoprotein fractions, revealed that CAD patients
had significantly more triglyceride and cholesterol in VLDL.
CAD patients had also less cholesterol in HDL but
triglyceride concentration in this lipoprotein fraction was
normal. LDL cholesterol and triglyceride concentrations
were similar in the CAD and in the normal groups. VLDL and
LDL apoB-100 and plasma apoA-I, apoA-II concentrations were
quantitated with specific radioimmunoassays (see Table II).
ApoB-100 concentration was increased in the VLDL but not in
the LDL of the CAD patients. In these patients, apoA-I and

TABLE IV

ApoB-100 Fractional Catabolic Rate (d^{-1}) and Production Rate (mg/d)

Patients	VLDL-apoB-100		LDL-apoB-100	
	FCR	PR	FCR	PR
Control (47-65; n=3)$	5.8 ± 1.1	791 ± 190	0.5 ± 0.1	786 ± 180
Young Volunteers (23-29; n=3)	5.6 ± 1.2	692 ± 130	0.4 ± 0.1	845 ± 208
CAD (48-69; n=5)	4.8 ± 0.8**	1368 ± 272**	0.6 ± 0.1**	1478 ± 204**
HTG (54-72; n=4)	2.8 ± 0.7**	1819 ± 486**	0.6 ± 0.1**	1218 ± 312**

$ In parenthesis the age range and the number of subjects

** $p < 0.05$ vs Control

apoA-II plasma concentrations were significantly decreased. In summary, CAD patients had increased VLDL levels as indicated by elevated cholesterol, triglyceride and apoB-100 levels in this fraction, whereas HDL where cholesterol, and apoA-I and apoA-II poor but with a normal triglyceride content. LDL were present in plasma in normal concentration and had normal composition.

Because of the limitation in the number of radioiodinated tracers that can be followed at one time, simultaneous determination of apoA-I, apoA-II and apoB-100 turnover rate in the volonteers was not possible. We thus resorted to initially investigate apoA-I and apoA-II metabolism in a group of CAD patients selected on the basis of low apoA-I plasma concentration and, upon demonstration of homogeneity in this group in terms of apoA-I and apoA-II metabolism we measured apoB-100 turnover rate in VLDL and LDL of other similarly hypoapoalfalipoproteinemic patients. In discussing the interrelationships between HDL apolipoprotein and apoB-100 metabolism in this subgroup of CAD patients, it was assumed that the second group of volonteers had the same defect in apoA-I and apoA-II metabolism that was leading to hypoapoalfalipoproteinemia the first group of patients.

The results on apoA-I and apoA-II metabolism in CAD patients with low apoA-I levels, have been presented elsewhere (31). In summary CAD patients had increased turnover of apoA-I and apoA-II but the synthetic rate of these apolipoproteins was normal. These data suggested that the primary metabolic defect leading to hypoapoalfalipoproteinemia is not at the level of HDL apolipoprotein synthesis. Rather it may originate from the abnormal composition of the HDL that in these patients are enriched in triglyceride. We then addressed the question of whether the metabolism of the apoB-100 containing lipoproteins is abnormal in CAD patients as this, as already discussed, may affect HDL concentration and composition. The abnormally elevated VLDL level in these patients was pointing out at this possibility.

Studies of apoB-100 metabolism in VLDL, IDL and LDL were undertaken in five normolipidemic subjects with CAD and decreased apoA-I plasma level and in three healthy volonteers in the same age group as the CAD patients. Moreover we studied apoB-100 metabolism in three other healthy volonteers in their twenties as well as in four hypertriglyceridemic subjects two of whom had CAD. This last group of patients was included since hypertriglyceridemia is almost invariably associated to defective VLDL apoB-100 metabolism, to decreased apoA-I and apoA-II plasma levels and to the presence of triglyceride-rich HDL (19,29). For these studies, I-125-VLDL and I-131-LDL were obtained by radioiodination of the recepient's own lipoproteins. The two tracers were injected simultaneously. Studies were carried out in a metabolic ward and the volonteers were put on a hypochaloric diet deficient in carbohydrates for the first two days following injection and then switched to an isocaloric diet for the remaining of the study period which ended at day ten. ApoB-100 specific activity in VLDL (d<1.006 g/ml), IDL (d=1.006-1.019 g/ml) and LDL (d=1.019-1.063) was determined by precipitating apoB-100 with isopropanol. Metabolic parameters such as the fractional catabolic rate and the production rate were calculated by analyzing the specific activity decay curves by manual peeling. In any case two exponentials were sufficient to describe the I-125-VLDL and I-131-LDL apoB-100 decay curves. The relevant clinical parameters of the subjects partecipating to the study and the turnover study results are reported in Table III and IV respectively. Overproduction of VLDL apoB-100 was observed in four of the five normolipidemic subjects with CAD and in all the patients with hypertriglyceridemia. In these patients VLDL apoB-100 daily production was in excess of 1000 mg. Abnormalities in apoB-100 metabolism also included decreased VLDL fractional catabolic rate consistent with a delay in conversion to lower density lipoproteins, and increased fractional catabolic rate of LDL. In the healthy volonteers as well as in the CAD patients, LDL apoB-100 production rate matched closely that of VLDL apoB-100. Assuming a simple precursor-product relationship between VLDL, IDL and LDL this indicate that in these patients most of VLDL are ultimately converted to LDL. By contrast in hypertriglyceridemic patients production rate of LDL apoB-100 was significantly less that of VLDL apoB-100 suggesting that a sizable fraction of VLDL is directly catabolized in the tissues prior to its conversion to LDL. Noteworhthy, LDL apoB-100 fractional catabolic rate in CAD and hypertriglyceridemic patients was similar and significantly faster than that measured in the plasma of the healthy volonteers.

THE METABOLIC BASIS OF NORMOLIPIDEMIC DYSLIPOPROTEINEMIA IN PATIENTS WITH CORONARY ATHEROSCLEROSIS

The finding that in CAD patients with low apoA-I levels, apoA-I and apoA-II synthetic rate is normal, whereas apoB-100 synthesis is increased strongly points out to an abnormality in apoB-100 gene regulation as the primary metabolic defect in these subjects. We have previously proposed (31), in discussing the mechanism leading to hypoapoalfalipoproteinemia in CAD pateints, that apoA-I and apoA-II rapid catabolism may be due to the abnormal composition of the HDL particles. Our group of CAD patients had in fact HDL enriched in triglycerides. The question arising is how delayed VLDL catabolism, accelerated HDL apoA-I and apoA-II and LDL apoB-100 catabolism and finally abnormal HDL composition may be integrated in a single model in which overproduction of apoB-100 is the primary defect.

With regard to VLDL delayed catabolism different possibilities may be considered. The first is that oversynthesis of VLDL, saturate VLDL catabolic pathway (32,33). Although plasma triglyceride are only slightly elevated, the number of VLDL particles may be significantly increased as we found more lipids and apoB-100 in this lipoprotein fraction. Because rate of conversion of VLDL to LDL ultimately depends on the number of particles processed in a given time, it is likely that rate of lipolysis depends not only on the total mass of triglyceride to be hydrolyzed but also on the number of VLDL particle to be processed. Based on the turnover data analysis, different Authors (34,35), have hypothesized the existence of an extravascular sequestered pool of VLDL remnants. Maybe the rate of VLDL to LDL conversion at this step is particle dependent. A second possibility, accounting for the delay in VLDL clearance, is that VLDL secreted by CAD subjects are compositionally abnormal. This is thought to occur in hypertriglyceridemic as well as in hyperapobetalipoproteinemic patients (20,36). An abnormal composition may retard VLDL particle processing for example as if they are poor substrate for lipoprotein-lipase. Moreover VLDL particles with abnormal composition may end their delipidization cycle well before reaching LDL status (37). At this stage they may be directly catabolized. Kasaniemi has previously reported that there is increased direct catabolism of VLDL in CAD patients (32).

Impaired removal of VLDL may directly or indirectly affect HDL metabolism in several way. There may be increased transfer of triglyceride from VLDL to HDL leading to the formation of HDL particles with abnormal composition. This mechanism has already been proposed to be responsible for the generation of triglyceride-rich HDL particles in hypertriglyceridemic patients (27,28). Another possibility is that normal flux of VLDL surface components to HDL is deficient - because of the impaired VLDL lipolysis - and does not match the HDL metabolic demand for these components (38). Moreover, since VLDL and intestinal chylomicron utilize the same delipidization pathway, there may be competition for lipoprotein-lipase and/or apolipoprotein cofactors. This also would affect HDL composition and catabolism as flux of chylomicron surface components appears necessary for HDL formation and maturation in plasma (26).

Hypercatabolism of LDL in conjunction with oversynthesis of VLDL has been reported by different Investigators (19). It has been observed in hypertriglyceridemic (39), in obese and CAD patients (32). LDL are eliminated from plasma either through a receptor dependent as well as independent pathway. The one that is receptor mediated, is responsible for most of the clearance of plasmatic LDL (40). The liver is the major site of catabolism of LDL (8). In this organ a B/E receptor is expressed that recognizes as ligand either apoB-100 and apoE. One possibility is that there is an increased number of B/E receptors available for LDL uptake. This may occur because more cholesterol is leaving the hepatocytes with VLDL and this loss must be compensated by increased return of LDL cholesterol. Thus the hepatocyte would respond by increasing the number of B/E receptors to promote enhanced influx of lipoproteins. This view is supported by the recent finding that in the hepatocyte, cholesterol synthesis, B/E receptor expression and apoB-100 synthesis are coordinated processes (42). Moreover different Investigators have shown that there is an inverse relationship between VLDL overproduction and LDL catabolism which may be secondary to enhanced LDL receptor expression (19,41). LDL hypercatabolism may also arise because an abnormal LDL composition is increasing the lipoprotein affinity for the B/E receptor (43,44). No data are yet available on receptor affinity of LDL isolated from CAD patients. We found that CAD subjects had normal levels of cholesterol, triglyceride and apoB-100 in LDL. The cholesterol to apoB-100 ratio was also normal in a group of normolipidemic CAD patients studied by Vega (45). On the other hand, Crouse et al (46) have found that in patients with CAD, LDL particles are smaller in size, but their chemical composition was not investigated.

In the above discussion primary deficiency of lipoprotein-lipase or defects of the B/E receptor activity, have not been considered. Prolonged residence in plasma of VLDL may arise if there is less lipoprotein-lipase available or the enzyme has low activity (47). There are reports that lipoprotein-lipase activity is decreased in CAD patients (48). In others a functional deficiency of hepatic lipoprotein-lipase has been observed (49). Hypercatabolism of LDL may be secondary to abnormalities in the B/E receptor. Either the liver receptors may be present in increased number or have increased affinity for LDL. Enhanced LDL uptake may increase VLDL production as more cholesterol is recirculating back to the liver. A number of normolipidemic patients with CAD may have these primary metabolic defects. In others, dyslipoproteinemia may be caused by apoB-100 mutants and so on. We have limited our discussion to consider VLDL oversynthesis considering the fact that it seems to occur frequently among CAD patients (32) and because is the primary metabolic defect in other dyslipoproteinemic conditions strongly associated with CAD (50).

CONCLUSIONS

Most subjects that develop CAD have total and LDL cholesterol levels that are below the 95% percentile (51). Epidemiological studies that have been carried out in these patients, invariably indicate that one or more apolipoprotein or lipoprotein lipid level is abnormal even when the individuals are normolipidemic (52). Virtually in any case there is decreased HDL cholesterol and/or apoA-I and apoA-II (52). In particular apoA-I measurements seems to possess greater predictive value for CAD than any other lipoprotein related variable (29). The results of the turnover study carried out in a group of hypoalfalipoproteinemic CAD patients, indicate that apoA-I and apoA-II are normally produced, whereas the metabolism of the triglyceride-rich lipoproteins is abnormal. We speculate that low HDL levels in these same patients is secondary to VLDL overproduction. Further studies on larger group of CAD patients including those with normal HDL levels are necessary to answer more definitely to the question of whether in normolipidemic CAD patients low HDL cholesterol or apoA-I concentration are always predictive of apoB-100 metabolic abnormalities. Nontheless the results obtained so far are of relevance in several ways. Firstly, additional circumstantial evidence is provided in favor of the view that total plasma lipid levels within the normal range, per se do not exclude abnormalities in lipoproteins metabolism. Alterations in plasma lipoprotein distribution and composition on the other hand may predict.them. Secondly, they raise the question of whether patients with normolipidemic dyslipoproteinemia are candidate to preventive (diet) and/or pharmacological intervention as has been recommended for hyperlipidemic patients. We believe so as the concept of total plasma lipid as risk factor has matured into one that considers critical how lipids are distributed in plasma and the role of abnormal and atherogenic lipoproteins (53). There is evidence that some hypolipidemic agents may normalize lipoprotein distribution and composition even in normolipidemic subjects (54). Finally the similarities in terms of metabolic abnormality in CAD and hypertriglyceridemic patients raises the possibility that patients with these diseases may share some similarity. It is likely that in some subjects normolipidemic dyslipoproteinemia is a permissive status for hyperlipoproteinemia which will become overt following dietary, hormonal and/or environmental challange.

REFERENCES

1) W.B. Kannel, W.P. Castelli and T. Gordon: Cholesterol in the prediction of atherosclerotic disease. New perspectives in the Framingham study. Ann. Intern. Med. 90:85 (1979)

2) Pooling Project Research Group: Relationship of blood pressure, serum cholesterol, smoking habit, relative weight and ECG abnormalities to incidence of major coronary events: final report of the pooling project. J. Chronic Dis. 31:201 (1978)

3)   J.D. Neaton, L.H. Kuller, D. Wentworth and N.O. Borhani: Total cardiovascular mortality in relation to cigarette smoking, serum cholesterol concentration, and diastolic blood pressure among black and white males followed up to five years. Am. Heart J. 108:759 (1984)

4)   Lipid Research Clinic Program: The lipid research clinics coronary primary prevention trial. I. Reduction in the incidence of coronary heart disease. JAMA 251:31 (1984)

5)   M.H. Frick, O. Elo, K. Haapa et al.: Helsinki Heart Study: Primary-prevention trial with gemfibrozil in middleaged men with dyslipidemia. New Engl. J. Med. 317:1237 (1987)

6)   D.H. Blankenhorn, S.A. Nesim, R.L. Johnson et al: Beneficial effect of combined colestipol-niacin therapy on coronary atherosclerosis and coronary venous bypass grafts. JAMA 257:3233 (1987)

7)   A.M. Gotto, Jr, H.J. Pownall and R.J. Havel: Introduction to the plasma lipoproteins. Methods Enzymol. 128:3 (1986)

8)   M.S. Brown and J.L. Goldstein: A receptor mediated pathway for cholesterol homeostasis. Science 232:34 (1986)

9)   G. Thompson: Apolipoproteins: Determinants of lipoprotein metabolism. Br. Heart J. 51:585 (1984)

10)   P. Avogaro, G. Cazzolato, G. Bittolo Bon, and G.B. Quinci. Are apolipoproteins better discriminators than lipids for atherosclerosis? Lancet I:901 (1979)

11)   A. Sniderman, S. Shapiro, D. Marpole et al: Association of coronary atherosclerosis with hyperapobetalipoproteinemia (increased protein but normal cholesterol levels in human plasma low density (beta) lipoproteins). Proc. Natl. Acad. Sci. USA 77:604 (1980)

12)   S.H. Chen, C.Y. Yang, P.F. Chen et al: The complete cDNA and amino acid sequence of human apolipoprotein B-100. J. Biol. Chem. 261:12918 (1986)

13)   T.J. Knott, S.C. Rall, T.L. Innerarity, et al: Human apolipoprotein B: Structure of carboxyl-terminal domains, site of gene expression, and chromosomal localization. Science 230:37 (1985)

14)   J.P. Kane, D.A. Hardman, and H.E. Paulus: Heterogeneity of apolipoprotein B: Isolation of a new species from human chylomicron. Proc. Natl. Acad. Sci. USA 77:2465 (1980)

15) R.J. Havel: Origin, metabolic fate and metabolic function of plasma lipoproteins. In: Contemporary Issues in Endocrinology and Metabolism. D. Steinberg and J.M. Olefsky eds. Churchill Livingstone Inc, New York pp 117-141 (1986)

16) C.Y. Yang, S-H. Chen, S.H. Gianturco, et al: Sequence, structure, receptor binding domains and internal repeats of human apolipoprotein B-100. Nature 323:738 (1986)

17) L.G. Fong, S. Parthasarathy, J.L. Witztum, and D. Steinberg: Nonenzymatic oxidative cleavage of peptide bonds in apoprotein B-100. J. Lipid Res. 28:1466 (1987)

18) M.E. Haberland, D. Fong, and L. Cheng: Malondialdehyde-altered apoB protein occurs in arterial lesion of Watanabe heritable hyperlipidemic rabbits. Arteriosclerosis 8:526A (1987)

19) Y.A. Kesaniemi, G.L. Vega and S.M. Grundy: Kinetics of apolipoprotein B in normal and hyperlipoproteinemic man: Review of current data. In: Lipoprotein Kinetic and modeling. M. Berman, S.M. Grundy and B.V. Howard eds. Academic Press, New York pp 182-207 (1982)

20) S.M. Grundy: Pathogenesis of hyperlipoproteinemia. J. Lipid Res. 25:1611 (1984)

21) R.J. Deckelbaum, E. Granot, Y, Oschry et al: Plasma triglyceride determines structure-composition in low and high density lipoproteins. Arteriosclerosis 4:226 (1984)

22) G. Utermann, U, Langenbeck, U. Beisiegel and W. Weber: Genetic of the apolipoprotein E system in man. Am. J. Hum. Genet. 32:339 (1980)

23) W.J. Schneider, P.T. Kovanen, M.S. Brown, et al: Abnormal binding of mutant apoprotein E to low density lipoprotein receptor of human fibroblasts and membranes from liver and adrenal of rats, rabbits and cows. J. Clin. Invest. 68:1075 (1981)

24) N. Fidge, P.J. Nestel, T. Ishikawa, et al: Turnover of apoproteins A-I and A-II of high density lipoproteins and the relationship to other lipoproteins in normal and hyperlipidemic individuals. Metabolism 29:643 (1980)

25) P. Magill, S.N. Rao, N.E. Miller, et al: Relationship between the metabolism of high density and very low density lipoproteins in man: Studies of apolipoprotein kinetics and adipose tissue lipoprotein lipase activity. Eur. J. Clin. Invest. 12:113 (1982)

26) S. Eisenberg: High density lipoprotein metabolism. J. Lipid Res. 25:1017 (1984)

27) L.F.B. Chang, G.J. Hopkins, and P.J. Barter: Particle size of high density lipoproteins as a function of plasma triglyceride concentration in human subjects. Atherosclerosis 56:61 (1985)

28) S. Eisenberg: Lipoprotein abnormalities in hypertriglyceridemia: Significance in atherosclerosis. Am. Heart J. 113:555 (1987)

29) N.E. Miller: Association of high density lipoprotein subclasses and apolipoproteins with ischemic heart disease and coronary atherosclerosis. Am. Heart J. 113:589 (1987)

30) Y.A. Kesaniemi, and S.M. Grundy: Overproduction of low density lipoproteins associated with coronary heart disease. Arteriosclerosis 3:40 (1983)

31) G. Ghiselli, J. Heibig, F. Turturro, et al: ApoA-I and apoA-II metabolism and coronary artery disease. In: Drugs Affecting Lipid Metabolism. R. Paoletti, D. Kritchevsky and W.L. Holmes eds. Springer-Verlag, Berlin pp 63-68 (1987)

32) Y.A. Kesaniemi, W.F. Beltz and S.M. Grundy: Comparison of metabolism of apolipoprotein B in normal subjects, obese patients, and patients with coronary artery disease. J. Clin. Invest. 76:586 (1985)

33) E.A. Nikkila and M. Kekki: Polymorphism of plasma triglyceride kinetics in normal human adult subjects. Acta Med. Scand. 190:49 (1971)

34) M. Berman, M. Hall, III, R.I. Levy, et al: Metabolism of apoB and apoC lipoproteins in man: Kinetic studies in normal and hyperlipoproteinemic subjects. J. Lipid Res. 19:38 (1978)

35) W.F. Beltz, Y.A. Kesaniemi, B.V. Howard and S.M. Grundy: Development of an integrated model for analysis of the kinetics of apolipoprotein B in plasma very low density lipoproteins, intermediate density lipoproteins, and low density lipoproteins. J. Clin. Invest. 76:575 (1985)

36) A.D. Sniderman, C. Wolfson, B. Teng, et al: Association of hyperapobetalipoproteinemia with endogenous hypertriglyceridemia and atherosclerosis. Ann. Int. Med. 97:833 (1982)

37) C.J. Packard, A.Munro, A.R. Lorimer, et al: Metabolism of apolipoprotein B in large, triglyceride-rich very low density lipoproteins of normal and hypertriglyceridemic subjects. J. Clin. Invest. 74:2178 (1984)

38) J.R. Patsch, A.M. Gotto, Jr, T, Olivecrona, and S. Eisenberg: Formation of high density lipoprotein-2 like particles during lipolysis of very low density lipoprotein in vitro. Proc. Natl. Acad. Sci. USA 75:4519 (1978)

39) G.L. Vega, W.F. Beltz, and S.M. Grundy: Low density lipoprotein metabolism in hypertriglyceridemic and normolipidemic patients with coronary artery disease. J. Lipid Res. 26:115 (1985)

40) Y.A. Kesaniemi, J.L. Witztum, and U.P. Steinbrecher: Receptor-mediated catabolism of low density lipoprotein in man: Quantitation using glucosylated low density lipoproteins. J. Clin. Invest. 71:950 (1983)

41) J.C. Monge, J.M. Hoeg, S.W. Law, et al: Human apolipoprotein B mRNA regulation: Role of apoB-containing particles and LDL receptor pathway. Arteriosclerosis 6:528A (1986)

42) H.N. Ginsberg, N-A. Le, and J.C. Gibson: Regulation of the production and catabolism of plasma low density lipoproteins in hypertriglyceridemic subjects. J. Clin. Invest. 75:614 (1985)

43) G.L. Vega, and S.M. Grundy: Mechanisms of primary hypercholesterolemia in humans. Am. Heart J. 113:493 (1987)

44) Y. Kleinman, S. Eisenberg, Y. Oschry, et al: Defective metabolism of hypertriglyceridemic low density lipoproteins in coltured human skin fibroblasts. Normalization with Bezafibrate therapy. J. Clin. Invest. 75:1796 (1985)

45) V.G. Vega, and S.M. Grundy: Comparison of apolipoprotein B to cholesterol in low density lipoproteins of patients with coronary heart disease. J. Lipid Res. 25:580 (1984)

46) J.R. Crouse, J.S. Parks, H.M. Schey, and F.R. Kahl: Studies of low density lipoprotein molecular weight in human beings with coronary artery disease. J. Lipid Res. 26:566 (1985)

47) A.P. Goldberg, A. Chait, and J.D. Brunzell: Postprandial adipose tissue lipoprotein lipase activity in primary hypertriglyceridemia. Metabolism 29:223 (1980)

48) J.D. Barth, H. Jensen, P.G. Hugenholtz, and J.C. Birkenhager: Post-heparin lipases, lipids and related hormones in men undergoing coronary artheriography to assess atherosclerosis: Atherosclerosis 48:235 (1983)

49) C. Breier, V. Muhlberger, H. Drexel, et al: Essential role of post-heparin lipoprotein lipase activity and of plasma testosterone in coronary artery disease. Lancet I:1242 (1985)

50) A. Chait, J.J. Albers, and J.D. Brunzell: Very low density lipoprotein overproduction in genetic forms of hypertriglyceridemia. Eur. J. Clin. Invest. 10:17 (1980)

51) W.P. Castelli, and K. Anderson: A population at risk. Prevalence of high cholesterol levels in hypertensive patients in the Framingham study. Am. J. Med. 80(2A):23 (1986)

52) P. Avogaro, G. Cazzolato, G. Bittolo Bon, et al: Lipoproteins derangement in human atherosclerosis. In: Lipoprotein and coronary atherosclerosis. G. Noseda, C. Fragiacomo, R. Fumagalli, and R. Paoletti eds. Elsevier Biomedical, Amsterdam pp 123-128 (1982)

53) R.J. Havel: Lowering cholesterol, 1988: Rationale, mechanisms, and means. J. Clin. Invest. 81:1653 (1988)

54) P. Moulin, M-C. Bourdillon, L. de-Parscau, et al: High density lipoprotein alterations induced by Bezafibrate in healthy male volonteers. Atherosclerosis 67:17 (1987)

CURRENT STATUS ON THE APO E-RECEPTOR

Beisiegel,U., Ihrke,G., Weber, W., Lohse,P. and Greten,H.,

Med. Kern- und Poliklinik, Universitätskrankenhaus

Eppendorf, Hamburg, FRG

INTRODUCTION

The potential apo E-receptor has become a special topic in the lipoprotein field. A receptor protein had been described, but it turned out to be much more complicated than expected. In this paper I will give you some evidence that an apo E receptor should exist and tell you about the history of the putative E-receptor. Last not least I will report on our current knowledge concerning apo E binding proteins.

EVIDENCE FOR THE PRESENCE OF AN APO E-RECEPTOR

It is known that there is a rather fast catabolism of chylomicron remnants (CMR) what indicates that a specific "remnant-receptor" is present in the liver (1,2). Several studies showed that apo E is responsible for the clearance of CMR in the liver (3,4). In Typ III patients a defect in the E-2 binding to the LDL-receptor causes a remnant accumulation. This gives evidence that the LDL-receptor is involved in CMR catabolism (5). However, the fact that in patients with familial hypercholesterolemia due to a LDL-receptor defect the clearance of CMR is normal.A second LDL-receptor independent mechanism has therefore to be postulated. This mechanism is proposed to be an unregulated receptor, not recognizing apo B (6). Both receptors seem to be genetically distinct and undergo independent control (7).

ISOLATION OF AN APO E-RECEPTOR

1986 Mahley et al published the isolation and characterization of an apolipoprotein E-receptor (8). This paper was taken as the proof for the concept of an apo E-receptor and seemed to clarify the nature of the receptor protein. To further characterize this receptor protein we started a collaboration with Dr. Mahleys group. We intended to produce a monoclonal antibody against this purified protein, to use it as a specific marker. After a sucessful fusion we had clones producing antibodies which reacted with the purified receptor in Western Blot analysis. While careful analysis of the specificity of the clones on discontinous SDS-PAGE (9) we detected that the monoclonal antibody recognized a protein with slightly higher molecular weight than the polyclonal antibody produced by Mahley's group. We estimated the molecular weight for the new protein with 59 kD compared to 56 kD for the published receptor. Both proteins were present in the fraction eluted from an apo E-affinity column, as described for the purification of the E-receptor.

Further characterization of the 56 kD E-receptor, initiated by these findings revealed in an amino acid sequence homology to the mitochondrial enzyme F1-ATpase. This intracellular enzyme is able to bind apo E with rather high affinity, but most probably it is not responsible for the CMR uptake in liver cells (10).

A SECOND APO E-BINDING PROTEIN

For the 59 kD protein which we detected with the monoclonal antibody in crude membrane fractions of human liver, we demonstrated E-binding capacity with affinity column and ligand blotting. We called it therefore an E-binding protein.

To further characterize this protein we used ligand blotting to be able to differentiate it from the LDL-receptor and from the ATpase. With a series of experiments we showed the following characteristics for the in vitro binding to the 59 KD protein:
The binding of apo E-liposomes and apo E containing lipoproteins to the 59 kD band was Calcium-independent. These findings is supported by the characterization of an Calcium-independent lipoprotein binding site described by Cooper et al (presentation on the AHA 1987), as well as by the data of Floren showing Calcium-independent binding of triglyceride-rich lipoproteins to human liver membranes. The binding to the 59 kD protein from human liver could be demonstrated with Western blotting for all three isoforms of apo E. In parallel experiments we could show apo E-2 binding to Hep G2 cells. The amount of apo E-2 binding was only slightly lower than apo E-3 and thereby considerably higher than expected for the LDL-receptor. – With these results we decided to consider this E-binding protein as a candidate for the potential apo E-receptor and initiated further studies.

VARIOUS PROTEINS WITH APO E-BINDING CAPACITY

Gudrun Ihrke, a graduate student from my laboratory had the chance to work in Prof. Stoffel's group for molecular biological studies on the E-binding protein. She screened a human liver c-DNA library with our monoclonal antibody and found two clones. The longer clone was partially sequenced, and reading basepairs for 206 amino acids they found a 95% homology with the protein disulphid isomerase (PDI), an enzyme in the ER (11). These result told us that there are at least two intracellular enzymes which are able to bind apo E in vitro with very similar molecular weights.
To clarify how many more proteins might hide in the one band on SDS-Page, described to be the purified E-receptor, we used 2-dimensional electrophoresis. The isoelectric focusing followed by SDS-PAGE revealed in 8-10 different spots with roughly the same molecular weight but different isoelectric points. That means we were facing the question of having 8-10 proteins able to bind to the apo E-affinity column. And it had to be evaluated which of those is found on the cell-surface, thereby beeing a candidate for the apo E-binding site.
One approach to this question was to use crosslinking reagents for coupling the apo E on the cell surface to its binding protein. We used the J-125 Denny Jaffe Reagent (12). Hep G2 cells were incubated with apo E-liposomes bound to the reagent. Apo E was than crosslinked by photoactivation. After chemical cleavage of the linkage we did indeed find a 59 kD band labelled on SDS-PAGE. This did confirm our concept of a 59 kD binding protein for apo E.
However, in addition to the 59 kD band we frequently saw another high molecular band labelled (around 500 kD). Further experiments have to show whether we look at a subunit of a large receptor complex or whether the high molecular weight band represents a heteromer or an aggregation of the 59 kD protein, beeing the binding protein itself.

Summary

There are several lines of evidence for an apo E mediated, receptor dependent uptake of CMR in the liver. The structure, however, of the potential receptor protein is not yet known. – So far we described E-binding proteins, some of which are found to be intracellular enzymes and thereby no candidates for a receptor protein, and others need to be characterized.

References

1. C.M. Arbeeny and V.A. Rifici, The Uptake of Chylomicron Remnants and Very Low Density Lipoprotein Remnants by Perfused Rat Liver, J.Biol.Chem. 259:9662 (1984)

2. M. Carrella and A.D. Cooper, High Affinity Binding of Chylomicron Remnants to Rat Liver Plasma Membranes, Proc.Natl.Acad.Sci.USA 76:338 (1979)

3. B. Sherrill, T.L. Innerarity and R.W. Mahley, Rapid Hepatic Clearance of Canine Lipoproteins Containing Only the E Apolipoprotein by a High Affinity Receptor, J.Biol.Chem. 255:1804 (1980)

4. A.D. Cooper, S.K. Erickson, R. Nutik and M.S. Shrewsbury, Characterization of Chylomicron Remnant Binding to Rat Liver Membrane J.Lipid Res. 23:42 (1982)

5. A.L. Jones, G.T. Hradek, C. Hornick, G. Renaud, E.E.T. Windler and R.J. Havel, Uptake and Processing of Remnants of Chylomicrons and Very Low Density Lipoproteins by Rat Liver, J.Lipid Res. 25:1151 (1984)

6. M.S. Brown and J.L. Goldstein, Lipoprotein Receptors in the Liver, J.Clin.Invest. 72:743 (1983)

7. J.M. Hoeg, S.J. Demosky, R.E. Gregg, E.J. Scheafer and H.B. Brewer, Distinct Hepatic Receptors for Low Density Lipoprotein and Apolipoprotein E in Human, Sience 227:759 (1985)

8. D.Y. Hui, W.J. Brecht, E.A. Hall, G. Friedman, T.L. Innerarity and R.W. Mahley, Isolation and Characterization of the Apolipoprotein E Receptor from Canine and Human Liver, J.Biol.Chem. 261:4256 (1986)

9. D.M. Neville jr., Molecular Weight Determination of Protein–Dodecyl Sulfate Complexes by Gel Electrophoresis in a Discontinous Buffer System, J.Biol. Chem. 246:6328 (1971)

10. U. Beisiegel, W. Weber, J.R. Havinga, G. Ihrke, D.Y. Hui, M.E. Wernette-Hammond, C.W. Turck, T.L. Innerarity and R.W. Mahley, Apolipoprotein E-Binding Proteins Isolated from Dog and Human Liver, Arteriosclerosis 8:288 (1988)

11. T. Pihlajaniemi, T. Helaakoski, K. Tasanen, R. Myllylä, M.-L. Huhtala, J. Koiva and K. Kivirikko, Molecular Cloning of the β–Subunit of Human Prolyl 4–Hydroxylase. This Subunit and Protein Disulphide Isomerase are Products of the Same Gene, EMBO J. 6:643 (1987)

12. C.L. Jaffe, H. Lis and N.Sharon, New Cleavabel Photoreactive Heterobifunctional Cross–Linking Reagents for Studying Membrane Organization, Biochemistry 19:4423 (1980)

DNA POLYMORPHISMS OF THE GLUCOSE TRANSPORTER GENE IN NON-INSULIN
DEPENDENT DIABETES MELLITUS (NIDDM)

S. Li and D.J. Galton

Medical Professorial Unit
St. Bartholomew's Hospital
London EC1A 7BE

INTRODUCTION

Inherited metabolic defects in non-insulin dependent diabetes
mellitus (NIDDM) underlying insulin resistance remain obscure. Strong
genetic determinants lead to more than 95% of identical twins being
concordant for the disease (1) whilst up to 30% of relatives of NIDDM
may have abnormal glucose tolerance (2). In diabetic pedigrees the
disease does not segregate in simple Mendelian proportions, suggesting
that it is unlikely to be monogenic.

Many cellular defects have been reported to occur in insulin-
sensitive cells of NIDDM subjects. Early reports described defects in
the activities of intracellular kinases, including phosphofructokinase
(3) pyruvate kinase (4) and more recently defects in the tyrosine
kinase activity of the B-subunit of the insulin receptor (5, 6).
Other defects have been reported of insulin binding to its cellular
receptor (7) and in the activation of glucose transport into cells by
insulin (8).

Of the many described cellular defects, the problem is to
distinguish inherited abnormalities from those secondary to the
metabolic disturbances in this disease. One approach is to study DNA
polymorphic markers close to candidate genes to see if such linkage
markers associate in diabetic populations or segregate in affected
individuals in diabetic pedigrees. If so, this is presumptive evidence
for an inherited defect within the vicinity of the candidate gene.

One such candidate gene is the glucose transporter located on
chromosome 1, p 33 (9). It codes for a membrane bound glycoprotein
of approximately molecular size of 55,000 daltons responsible for the
transport of glucose into cells that is stimulated by insulin in some
target tissues (10). The reduced sensitivity of target cells to the
action of insulin could reside in the insulin receptor or in post
receptor response of proteins such as the glucose transporter or intracellular kinases. We here report a DNA polymorphic marker adjacent to
the glucose transporter gene that appears to associate in NIDDM subjects
of Caucasian origins.

PATIENTS AND METHODS

Fifty-eight unrelated caucasians (29 males), attending the diabetic
clinic at St Bartholomew's Hospital were studied. Non-insulin dependent
diabetes (NIDDM) was diagnosed according to clinical characteristics as
defined by WHO (11). Any case of secondary diabetes was excluded.

Sixty-six race and age-matched controls were recruited from a health
screening clinic, on the basis of a negative personal and detailed
family history of diabetes mellitus and normal fasting blood glucose
(fasting blood glucose <6.0 mmols/l). Blood glucose estimations were
made on fluoridated whole blood samples by a glucose oxidase method
('Technicon AA II' single channel analyser).

DNA Analysis

DNA was precipitated from 10 mls of whole blood by the method of
Kunkel et al (12). At least 10 ug of redissolved genomic DNA was
digested with the restriction endonucleases Xba-1 and Bgl II (Bethesda
Research Laboratories, Cambridge) according to manufacturers
instructions. The resulting fragments were separated according to size
by 0.85% agarose gel electrophoresis, denatured and transferred to
nitrocellulose filters (Schleider and Schull, Passel, West Germany)
by Southern blotting. Each filter was subsequently hybridised with a
32P-labelled glucose transporter cDNA sub-clone (hGT2-2) extending
from base pair - 200 to 281 (10). Hybridisation bands were
visualised by autoradiography at -70C using preflashed Amersham
Hyperfilm TM and intensifying screens.

Results

Digestion of leucocyte DNA with Xba-1 reveals a polymorphism,
probably due to a point mutation, yielding restriction fragments of
6.5 Kb (the XI allele) or 6.3 Kb (the X2 allele). These fragment
sizes were estimated by running Hind III digests of lambda phage
concurrently. The number of diabetic and control subjects with each
allele is shown in Table I. The proportion of subjects possessing the
X1 allele was significantly greater in the NIDDM group than controls
and the XI allele frequency was increased in diabetics compared to
controls (p <0.01).

TABLE I

GLUCOSE TRANSPORTER GENOTYPES : NUMBERS OF
CAUCASIANS POSSESSING ALLELES REVEALED BY Xba DIGESTION
OF LEUCOCYTE DNA

	n	Genotype Frequencies (%)			Allele Frequencies	
		X1.X1	X2.X2	X2.X2	X.1	X.2
Controls	66	3(4)	29(44)	34(52)	0.26	0.74
Non-insulin dependent diabetes	58	7(12)	37(64)	14(24)	0.44*	0.56

* p<0.01 compared in a 2 x 2 table by chi-square (= 8.3)

Discussion

Identification of individuals at risk from NIDDM at an early age might allow the introduction of preventive measures. At present, abnormalities of glucose tolerance are used as the diagnostic criteria for diabetes but are variably affected by age and past nutritional habits of the individual. In contrast, genetic markers have the potential to identify susceptible individuals from early childhood.

Using a cDNA subclone of the glucose transporter gene (covering nucleotides - 200 - 281) we have studied an Xba-1 restriction fragment length polymorphism in a caucasoid diabetic population.

An increased frequency of the XI allele was observed between NIDDM and control subjects ($p < 0.01$). The association can be observed in diabetic populations and suggests that the XI allele is probably within 0.1 - 0.5% recombination distance from a putative diabetogenic locus (13). Mapping experiments will be required to identify such a locus, and if it lies within the exon sequence of the glucose transporter gene, it may possibly constitute a major genetic determinant for the disease. Alternatively it may be in linkage disequilibrium with other abnormalities adjacent to the glucose transporter gene that impair carbohydrate metabolism. However even at this stage, it is apparent that DNA polymorphisms of the glucose transporter gene will be useful to the investigation of the inherited defects of diabetes mellitus.

ACKNOWLEDGEMENTS

The authors are grateful for financial support from the Wellcome Trust (DJG) and the British Diabetic Association (SL). We would like to thank the staff of the BUPA Medical Centre and Dr Edwin Gale of St Bartholomew's Hospital for permission to study diabetic patients under his care. The glucose transporter gene probe was a kind gift from Dr M Mueckler, Washington University School of Medicine, Missouri, U.S.A.

REFERENCES

1.   Barnett AH, Eff C, Leslie RDG, Pyke DA (1979). Diabetes in identical twins : a study of 200 pairs. Diabetologia 17 : 333 - 343.

2.   Kobberling J (1971). Studies on the genetic heterogeniety of diabetes mellitus. Diabetologia 7 : 46 - 49.

3.   Galton DJ, Wilson JPD (1971). The effect of starvation and diabetes on glycolytic enzymes in human adipose tissue. Clinical Science 41 : 543 - 553.

4.   Samian R, Stansbie D, Dawson A, Galton DJ (1980). Effects of alanine and fructose 16-bisphosphate on the activity of human adipose tissue pyruvate kinase. Clinical Science 59 : 1 - P.

5.   Freidenberg GR, Henry RR, Klein HH, Reichart DR, Olefsky JM (1987). Decreased kinase activity of insulin receptors from adipocytes of non-insulin dependent diabetic subjects. J Clin Invest 79 : 240 - 250.

6.   Comi RJ, Grunberger G, Gorden P, (1987).   Relationship of
     insulin binding and insulin   stimulated tyrosine kinase
     activity is altered in Type II Diabetes.   J. Clin. Invest.
     79 : 453 - 462.

7.   Olefsky JM, Reaven GM (1974).   Decreased insulin binding to
     lymphocytes from diabetic patients.   J Clin Invest 54   :   1323 -
     1328.

8.   Kashiwagi A, Verso MA, Andrews J, Vasquez B, Reaven GM,
     Foley JE (1983).   In vitro insulin resistance of human adipocytes
     isolated from subjects with non-insulin-dependent diabetes
     mellitus.   J Clin Invest 72   :   1246 - 1254.

9.   Shows TB, Eddy RL, Byers MG, Fukushima Y, Dehaven CR, Murray JC,
     Bell GT (1987).   Polymorphic human glucose transporter gene is
     on chromosome 1 p 31.3 - p.35   Diabetes 36   :   546 - 549.

10. Mueckler M, Caruso C, Baldrin SA, Panico M, Blench I,
     Morris HR, Allard WJ, Lienhard GE, Lodish HF (1985)
     Science 229 : 941 - 945.

11. World Health Organisation Expert Committee on Diabetes
     Mellitus, Second Report;   1980, Wtto   Technical Report Series
     646.

12. Kunkel LM, Smith KD, Boyer SH (1977).   Analysis of human Y
     chromosome specific reiterated DNA in chromosome variants.
     Proc. Natl. Acad. Sci USA.   74   :   124   - 1249.

13. Brock G, Collins GM ed (1987).   Molecular approaches to
     human polygenic disease;   Ciba Foundation Symposium 130.
     publ.   John Wiley & Sons.

COMPUTER-MODELLING OF HUMAN APOLIPOPROTEINS AND OF THEIR MUTANTS

M.Rosseneu[1], J.M. Ruysschaert[2], M. Froeyen[1]
and R.Brasseur[2]

(1) Dept.Clinical Biochemistry A.Z.St-Jan, Brugge
(2) Free University Brussels, Brussels (Belgium)

INTRODUCTION

Since the isolation and sequencing of the apolipoproteins, several attempts have been made towards the identification of the structural domains accounting for the specific properties of these proteins (1). Apolipoproteins are responsible for the transport of apolar lipid molecules in plasma. Their lipid-binding properties have been attributed to the amphipathic helices which are present in all of the apolipoproteins (2). The sequences of these proteins have a high degree of internal homology and the various apolipoproteins, including apo B share homologous domains with each other (3).

Several apolipoprotein mutants have been identified mainly for apo AI and apo E (4). The mutations seem to affect mostly the receptor-binding properties and to a lesser extent the lipid-binding function (5).

The use of physico-chemical techniques for the characterization of the mutants has not provided substantial information concerning the structure of these proteins, since most mutations involve only one amino acid substitution (4). We have chosen a more theoretical approach based upon the analysis of the sequence, the prediction of the structural properties of the helical segments to attempt a description at the molecular level of the organisation of the protein and the lipid in the lipid-apoprotein complexes.

The data obtained with apo AI and apo AI Milano and with apo A-IV are summarized in this paper.

1. CONFORMATION OF NORMAL APO AI AND OF APO AI Milano

The Arginine-Cysteine mutation at position 173 of apo AI in the Milano variant affects the HDL levels in the probands. One could therefore expect differences in the lipid-binding properties of these two apo AI isoforms. "In vitro" studies on the formation of phospholipid-apo AI complexes could not detect significant differences in the behaviour of the two isoforms (6). The physico-chemical properties of apo AI Normal and Milano are also very close.

We investigated the effect of the Arg173-Cys mutation, occuring in the center of the polar face of the amphipatic segment spanning residues 165 to 197 upon the properties of this helix. The mean hydrophobicity is significantly different for the Normal than for the Milano variant, -0.29 compared to -0.18, using the hydrophobicity scale of Eisenberg (7), though the amphipathic character of the helix is preserved.

Using a computer-modelling procedure previously described for amphipathic and transmembrane helices (8), the helix at residues 165-197 was oriented at the air-lipid interface, then surrounded with phospholipid molecules and finally the energy of interaction of this peptide-lipid complex was calculated.

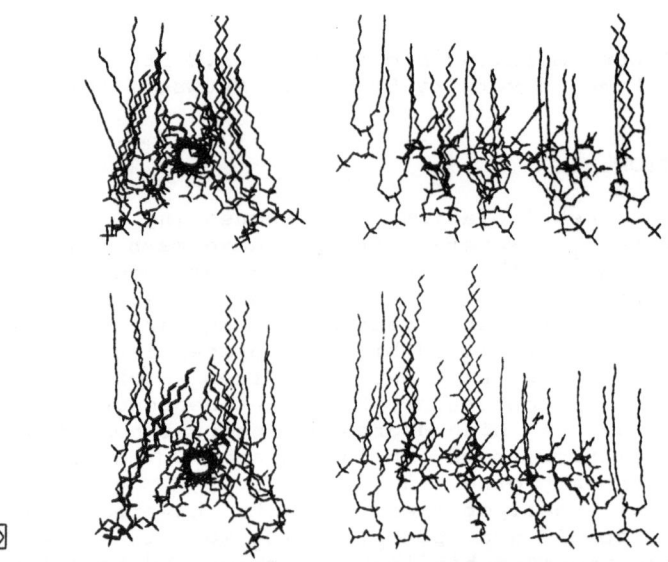

Fig.1. Skeleton representation of the helix 165-197 in the Normal apo AI (upper part) and in the Milano apo AI (lower part). The helix is represented perpendicular (left) and parrallel (right) to its axis, surrounded with phospholipids.

As shown in Fig.1, this particular helix was oriented parallel to the air-water inerface, as previously shown for other apo AI segments (8). The energies of interaction of the two helices with the lipids are very close (-12.0 and 11.8 Kcal/mol lipid for Normal and mutant apo AI respectively) in agreement with the experimental lipid-binding data (6). The introduction of a Cys residue at position 173 induces however a larger destabilization of the phospholipid acyl chains, whose orientation is more perturbed in the apo AI-Milano lipid complex than with the Normal apoprotein. Moreover 11 phospholipid molecules are required to surround the helix in the apo AI Milano compared to 10 in the Normal protein.

2. ASSEMBLY OF THE HELICAL SEGMENTS OF APO A-IV WITH LIPIDS.

Apo A-IV is the apolipoprotein with the highest degree of internal homology as 12 repeats of 22 residues, with amphipathic helical properties have been identified (Table 1).

Table 1. Helical amphipathic sgments and consensus sequence for human apo AIV. The segments are identified by their first residue.

13	D	Y	F	S	Q	L	S	N	N	A	K	E	A	V	E	H	L	Q	K	S	E	L
62	P	F	A	T	E	L	H	E	R	L	A	K	D	S	E	K	L	K	E	E	I	G
95	P	H	A	N	E	V	S	Q	K	I	G	D	N	L	R	E	L	Q	Q	R	L	E
117	P	Y	A	D	Q	L	R	T	Q	V	N	T	Q	A	E	Q	L	R	R	Q	L	T
139	P	Y	A	Q	R	M	E	R	V	L	R	E	N	A	D	S	L	Q	A	S	L	R
161	P	H	A	D	E	L	K	A	K	I	D	Q	N	V	E	E	L	K	G	R	L	T
183	P	Y	A	D	E	F	K	V	K	I	D	Q	T	V	E	E	L	R	R	S	L	A
205	P	Y	A	Q	D	T	Q	E	K	L	N	H	Q	L	E	G	L	T	F	Q	M	K
227	K	N	A	E	E	L	K	A	R	I	S	A	S	A	E	E	L	R	Q	R	L	A
249	P	L	A	E	D	V	R	G	N	L	R	G	N	T	E	G	L	Q	K	S	L	A
289	P	Y	G	E	N	F	N	K	A	L	V	Q	Q	N	E	Q	L	R	T	K	L	G
311	P	H	A	G	D	V	E	G	H	L	S	F	L	E	K	D	L	R	D	K	V	N
CONSENSUS	P	Y	A	D	E	L	K	A	K	L	S	Q	N.A	E	E	L	R	Q	S	L	A	

Using a three-dimensional computer modelling procedure (9), we assembled these segments with phospholipids while taking into account the mutual interactions between helices. This type of analysis was first developed for the assembly of hydrophobic helices spanning the phospholipid bilayer and can be applied to the assembly of the apolipoprotein amphipathic segments.

Hydrophobicity profiles were calculated in order to identify the most probable associations between the 12 amphipathic helices, based upon profile similarities. For each peptide, the hydrophobic and hydrophilic transfer energy along the alpha-helix axis were obtained by moving a 6 A window along the axis of the alpha-helix and calculating the sum of the respective transfer energies for all atoms included in the window. A cubic spline function was calculated yielding the hydrophobic and hydrophilic transfer energy along the alpha-helix.

In apo A-IV, the helical repeats 3-10 occur without interruption from residues 95 up to 261. These segments are separated by beta-turns including the two first residues of the next helix (Pro-X, except for Lys-X in residue 227-8) and the two last residues of the previous segments. These helices have to be located close to each other in view of steric considerations. Based on similarities of the hydrophobicity profiles, the remaining segments could be grouped in pairs : 1-10, 2-12, 3-11. These pairs were fit together with segments 3 to 10, which have to be in close vicinity, in the model shown in Fig 2. The central part of this model, consists of a phospholipid bilayer and the apo A-IV helices are oriented parallel to the phospholipid acyl chains, as in the bicycle-tyre model proposed by Segrest et al (2).

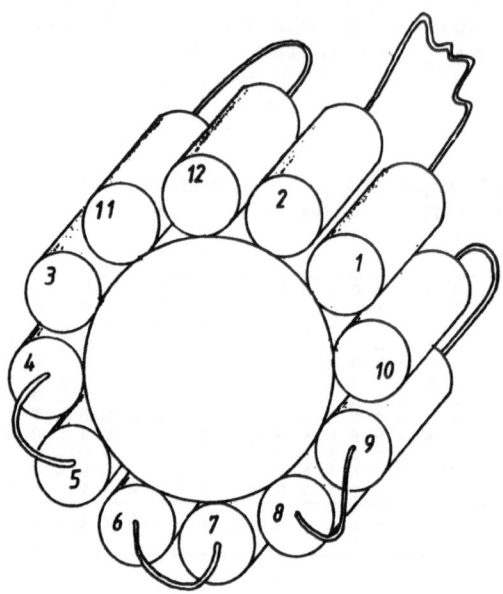

Fig.2. Schematic representation of the assembly of the apo AIV helices in an apo AIV-phospholipid complex.

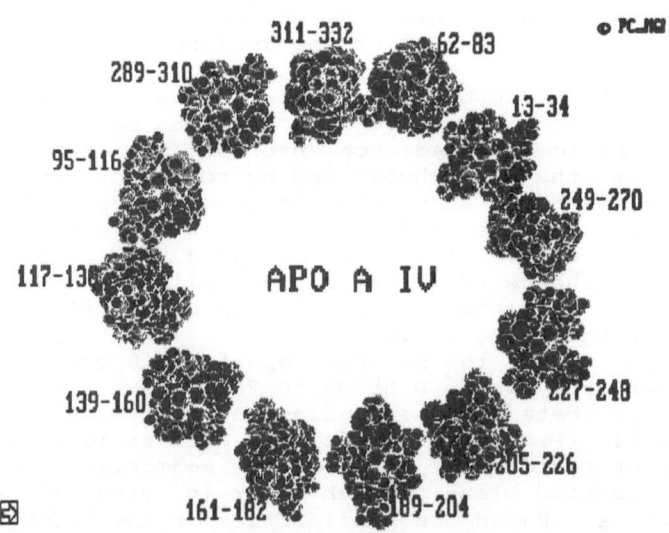

Fig.3. Computer drawing showing the orientation and assembly of the apo AIV helices in an apoprotein-lipid complex. Dark areas represent the hydrophobic residues.

The apolar face of the helices is directed towards the phospholipid acyl chains, while the polar faces are oriented towards water, shielding the hydrophobic residues from the aqueous environment. The phospholipid polar head groups are located towards the upper and lower surfaces of the cylinder. The length of an amphipathic segment (18 residues, not including the beta-turn) corresponds roughly to the length of the hydrophobic portion of the phospholipid bilayer. The distance between helix 2 and 3 and between 10 and 11 is spanned by segments of respectively 13 and 19 residues. These segments are enriched in polar residues (Glu, Arg) and might represent regions of the protein susceptible to fulfill a receptor-binding function or to possess antigenic properties.

The computer modelling programs described above for the calculation of the minimal energy of the lipid-protein complex were applied to this system and yielded the model depicted in Fig.3. In this model, the hydrophobic faces of the helices are drawn in black and all oriented towards the central lipid core, which is not represented. The hydrophilic faces (light) point towards the outside. In this model, the helices interact both with each other and with the lipids. The dimensions calculated from this model yield an outer diameter of 83 A and an inner diameter of 53 A. Based upon the surface of a phospholipid molecule spread at an interface, this complex could accommodate about 2X 35 phospholipid molecules /apo A-IV molecule. These theoretical results on the complex composition and size are in good agreement with the experimental data obtained by Rifici et al.(10) on the rat apo A-IV-DMPC complexes.

CONCLUSIONS

The analysis of the apolipoprotein sequences by the available prediction methods has enabled the identification of the candidate amphipathic helical segments and the characterization of their hydrophobicity and hydrophobic moment. Computer procedures, especially designed for the modelling of protein domains in association with lipids, have yielded the most probable orientation and interaction energies for the complexes. Finally, the calculation of the hydrophobicity profiles along the axis of the amphipathic helices enabled the assembly of the helical repeats of apo A-IV in a lipid-protein complex. Further experimental tests on the validity on this model, especially by studying the susceptibility of some segments to enzymatic degradation, are currently under progress.

REFERENCES

1. W-H Li, M.Tanimura, C-C luo, S.Datta and L.Chan, The apolipoprotein multigene family: biosynthesis, structure, structure-function relationships and evolution. J.Lipid Res. 29:245 (1988).

2.   J.P. Segrest, R.L. Jackson, J.D. Morrisett and A.M. Gotto, A molecular theory of lipid-protein interactions in the plasma lipoproteins. FEBS Lett. 38:247 (1974)

3.   H.De Loof, M.Rosseneu, C-Y.Yang, W-H.Li, A.M.Gotto and L.Chan. Apolipoprotein B: analysis of internal repeats and homology with other apolipoproteins. J.Lipid Res. 28: 1455 (1987).

4. J. Breslow, Apolipoprotein genetic variation and human disease. Phys. Reviews. 68 : 85 (1988)

5.   H.De Loof, M.Rosseneu, R.Brasseur, and J.M.Ruysschaert. Hydrophobicity profiles for detection of receptor binding domains on apolipoprotein E and the low density lipoprotein apolipoprotein(B-E) receptor. Proc.Natl.Acad.Sci.USA. 83: 2295 (1986)

6. G. Franceschini, G. Vecchio, G. Gianfranceschi, D. Magni, and C. Sirtori. Apolioprotein AI Milano. J. Biol. Chem. 260:16321 (1985).

7. D. Eisenberg, Three-dimensional structure of membrane and surface proteins. Annu.Rev.Biochem.53:595 (1984).

8.   R. Brasseur, H. De Loof, J.M. Ruysschaert and M. Rosseneu. Conformational analysis of lipid-associating proteins in a lipid environment. Biochim. Biophys. Acta. Accepted.

9.   R.Brasseur, Calculation of the three-dimensional structure of saccharomyces cerevisiae cytochrome b inserted in a lipid matrix. J.Biol.Chem.Accepted.

10.   V.A. Rifici, H.A. Eder and J.B. Swaney. Isolation and lipid-binding properties of rat apolipoprotein AIV. Biochim.Biophys.Acta 834 : 205 (1985).

MONOCLONAL ANTIBODIES TO HUMAN APOLIPOPROTEIN A-I

Alberico L. Catapano and Santica Marcovina

Institute of Pharmacological Sciences, University
of Milano and Scientific Institute San Raffaele
Milano, Italy

INTRODUCTION

Human plasma high density lipoprotein (HDL) contains approximatively 45% protein and 65% lipid by weight. Among lipids cholesteryl esters and phospholipids are major constituent, apolipoprotein A-I and apolipoprotein A-II account for 60 and 30% about of the HDL protein mass (1). Apolipoprotein A-I is synthetized as pre-proapoprotein; the pre-peptide is cleaved cotranslationally, while the proapo A-I is apparently secreted as such mainly by the intestine and liver (2). Conversion to the mature form (243 aminoacids) occurs in plasma via a specific enzyme that cleaves the pro (6 aminoacid) segment.

The apoprotein is a cofactor for the enzyme lecithin cholesterol acyl-transferase; the key enzyme for esterification of cholesterol in plasma (3). Recent work also points to a role of apo A-I in interacting with a specific cellular "receptor" for HDL, that might play a relevant role in the unloading of cholesterol from lipid-laden cells (4).

One of the approaches to the understanding of apolipoprotein A-I function might be the use of monoclonal antibodies to this apoprotein. Several examples for the use of monoclonal antibodies in mapping functionally relevant areas of enzymes, proteins, and apolipoproteins exist. Monoclonal antibodies to apolipoprotein B, in association with biochemical data, have helped in the identification of the apo B region (5) involved in the interaction with the LDL receptor, near the T3/T2 junction in apolipoprotein B-100. Monoclonal antibodies may also shed light on the conformational changes occurring during the interaction of the apolipoproteins with phospholipids: epitopes that are not expressed in the delipidated form may become evident upon reassociation with lipids, and vice versa.

Here we describe the production and characterization of six monoclonal antibodies to apolipoprotein A-I, and in particular the expression of their antigenic sites on Apo A-I before and after association with lipids.

MATERIALS AND METHODS

Human high density lipoprotein was isolated as HDL_2 and HDL_3 between densities 1.063-1.125 g/ml and 1.125-1.210 g/ml respectively, at 40.000 rpm for 48 hr at 12°C in a Beckman ultracentrifuge (6). The apolipoprotein A-I was isolated by column chromatography of HDL apoproteins. Appropriate fractions were collected and their purity determined by SDS gel electrophoresis. Complexes between purified Apo A-I and phosphatidyl coline were prepared as previously described (7); shortly egg PC and cholesterol were solubilized in chloroform-methanol (2:1), added to a disposable test tube, and dried under a steam of nitrogen. Apo A-I in 0.15 mM NaCl, 2 mM EDTA pH 7.4 was then added and the mixture vortexed for 30 s (Molar ratio to PC 1:100). Sodium cholate was added under continuous stirring to reach a final concentration of 4%. Cholate was removed by dialysis and the complex stored at 4°C under nitrogen until use.

Monoclonal antibodies were produced as described previously (8). Shortly male Balb/c mice were immunized with purified Apo A-I. The mouse with the highest specific antibody titre, as determined by enzyme-linked immunoassay, was boosted with intravenous injections of 50 ug of Apo A-I without any adjuvant for three consecutive days before the fusion. Splenocytes from the immunized mouse were fused with the mouse hybridoma cell line SP2/0-Ag14, and cultured in complete Iscove's medium. As cell proliferation approached confluency, supernatants were screened for the presence of specific antibodies by an enzyme-linked immunoassay as previously described.

The cells from the most positive wells were cloned by limiting dilution in 96 well microculture plates, yielding 26 clones that were recloned twice. Six stable clones showing the highest antibody titre to both purified Apo A-I and HDL were selected to be characterized afted being further subcloned three times. To obtain a large amount of antibodies, the hybridoma cells were injected into the peritoneal cavity of "pristane primed" Balb/c mice. The ascitic fluid were collected 7-14 days later and centrifuged. The monoclonal antibodies were purified from ascitic fluids by adsorption to Protein A-Sepharose and stored at -80°C.

HDL_2 and HDL_3 subfraction were labeled with ^{125}I using the procedure described by Bilheimer et al. (9). Free ^{125}I was removed by gel filtration on a Sephadex PD25 column. Labeled lipoprotein was dialyzed extensively at 4°C against PBS containing 0.01% Na_2EDTA. The ^{125}I precipitable with 10% trichloroacetic acid (TCA) was always more than 98%. The lipid-associated radioactivity ranged from 3 to 8% of TCA precipitable radioactivity, specific activity was 300-400 cpm/ng of protein. Apo A-I was also labeled with this procedure to a specific activity of 400-500 cpm/ng. TCA precipitable radioactivity was always greater than 98%. Both labeled HDL and Apo A-I were stored at 4°C under sterile conditions, and used within a week.

Fluid phase radioimmunoassay was performed in duplicate. Briefly, 100 ul of serial dilution of monoclonal antibodies were incubated with 100 ul of ^{125}I-labeled HDL (30,000 cpm, containing about 180 ng of protein). The assay buffer contained 1% bovine serum albumin, 0.02 M Tris-HCl, 0.1 M NaCl, 0.01% NaN_3 pH 7.4. After overnight incubation at 4°C, 100 ul of rabbit antimouse IgG were added (diluted to give a slight antibody excess), and the tubes incubated for 3 hr at room temperature. At the end of the incubation, 100 ul of standardized Pansorbin cells (Calbiochem-Behring), were added and let incubate 30 min. Alternatively, a one step procedure was used in which the rabbit anti-mouse IgG (diluted 1:50) were preincubated for 24 hr at 4°C with Pansorbin cells (1 ml of cell suspension per 10 ml of diluted antibody) prior the use in the RIA assay. Rabbit antimouse IgG/Pansorbin cells (100 ul) was added to the tubes and incubated for 6 hr at room temperature. The tubes were then washed with 2 ml of cold buffer and centrifuged (2000 x g, 30 min). Supernatants were removed and the radioactivity of the pellet was determined. Maximum precipitable radioactivity was determined by replacing the rabbit anti-mouse IgG with trichloroacetic acid (220 g/1). The non specific binding was determined by replacing the specific monoclonal antibodies with supernatant fluid from an irrelevant hybridoma.

For calculations, the percent of total ^{125}I-labeled HDL bound was expressed by B_0/T, where B_0 is the total ^{125}I-labeled HDL bound to antibodies, minus nonspecific binding, and T was the maximum trichloroacetic acid-precipitable radioactivity. Among the twenty-six hybridomas that secreted antibodies specific for human plasma Apo A-I six clones were selected by enzyme immunoassay on the basis of their capacity to bind equally to freshly purified Apo A-I and freshly isolated HDL on solid-phase. The details of the characterization of the six monoclonal antibodies used in this study are shown in **Table 1**.

TABLE 1. Monoclonal antibodies to human apolipoprotein A-I

Immunizing antigen	Hybridoma cell line	Hybridoma antibody	Monoclonal antibody	Antibody Chain type	
apo A-I	SP2/0-Ag 14	9-B2-D2	A-I- 9	IgG1	k
"	"	12-C4-B1	A-I-12	IgG2a	k
"	"	15-C4-H1	A-1-15	IgG2a	k
"	"	16-C6-C4	A-I-16	IgG1	k
"	"	19-D4-B6	A-I-19	IgG2b	k
"	"	57-H1-H9	A-I-57	IgG1	k

RESULTS

Preliminary experiments on the immunoprecipitation of ^{125}I-HDL in liquid phase RIA showed that the binding of the six monoclonal antibodies (measured at a final concentration of 80 ng of ^{125}I-HDL) varied from 55% for the clone A-I-16 to about 90% for all the other clones. To verify whether these data reflected the true availability of the epitopes at the HDL surface, or the percent of HDL binding could be improved by optimization of the assay conditions, we used rabbit anti-mouse IgG preincubated with Pansorbin cells instead of the two-step procedure. This approach did not modify the maximal binding of monoclonal antibody A-I-16 and A-I-9 which remained 55 and 90% respectively. Antibodies A-I-12, A-I-15,

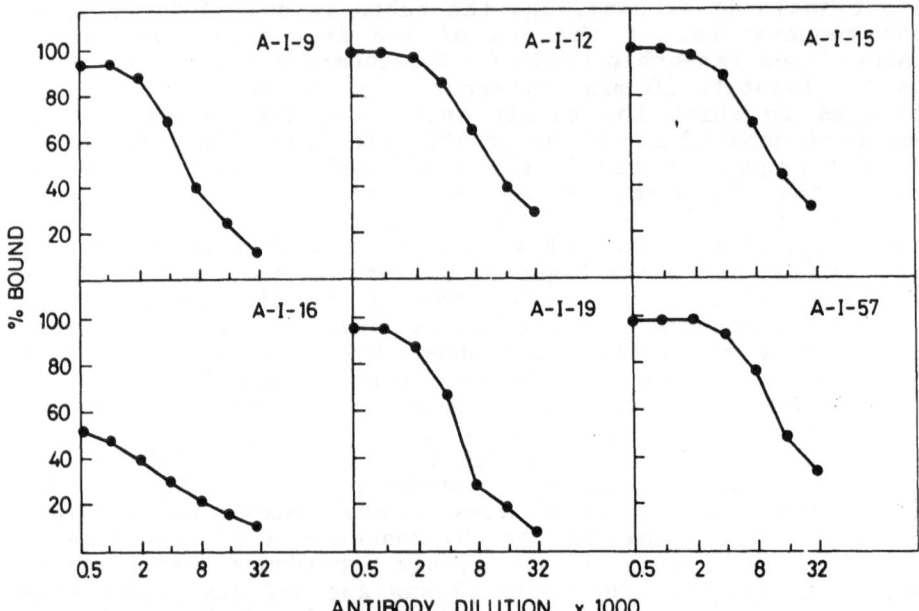

Figure 1. Immunoprecipitation of ^{125}I-labeled HDL by monoclonal antibodies. The percentage of trichloroacetic acid-precipitable radioactivity bound by each antibody is shown as a function of antibody diluition.

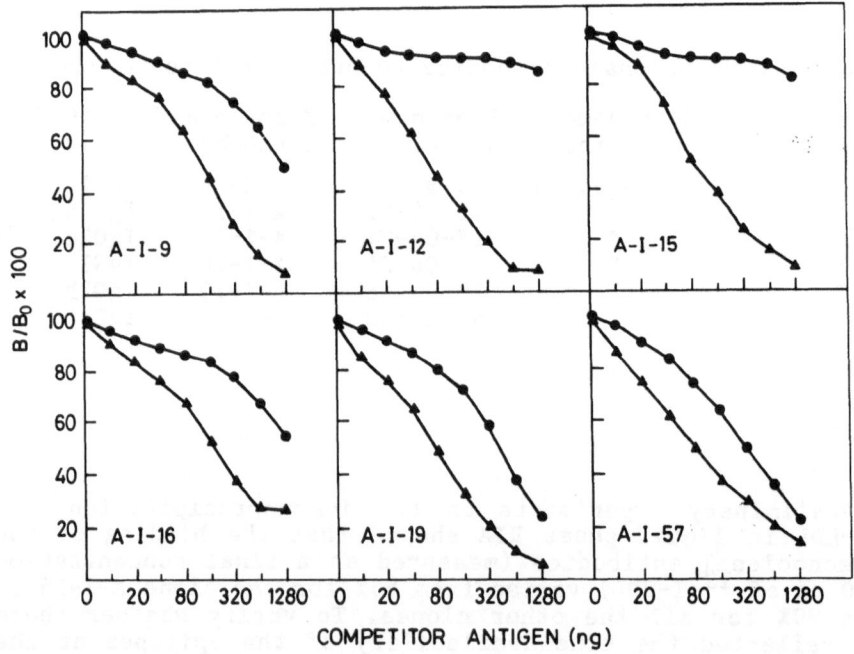

Figure 2. Displacement curves of ^{125}I-labeled HDL by HDL (▲) and purified apo A-I (●). Bound HDL was determined by incubating 100 ul of ^{125}I-labeled HDL (80 ng of protein) with 100 ul of competing antigen (HDL or apo A-I) ranging from 10 to 1280 ng, and 100 ul of monoclonal antibody at the diluition required for about 50% maximum binding.

and A-I-57, however, precipitated the totality of ^{125}I-labeled HDL, and the binding of clone A-I-19, was 96% of total HDL (**Fig. 1**).

To measure the antibody affinities of apolipoprotein A-I in HDL and soluble Apo A-I, a fluid phase radioimmunoassay was performed in which the ability of HDL and purified Apo A-I to compete for binding of ^{125}I-HDL was analyzed. Full displacement of ^{125}I-HDL by HDL was obtained with each monoclonal antibody but A-I-16.

An incomplete displacement of ^{125}I-HDL by purified Apo A-I was obtained with all monoclonal antibodies but A-I-57, and as shown in Fig. 2, the immunoactivities were less for free Apo A-I than for Apo A-I organized in HDL. Monoclonal antibodies A-I-12 and 15 particularly seem to be directed to epitope(s) very sensitive to the changes to which Apo A-I undergoes either due to the lipid dissociation or to the delipidation and storage. To further investigate the role of lipids in mantaining the antigenic structure of Apo A-I, the ability of Apo A-I vesicles to compete for binding of ^{125}I-HDL was analyzed in a fluid phase radioimmunoassay. We compared the displacement curves obtained by monoclonal antibody A-I-12, and A-I-15, which showed almost no displacement with purified Apo A-I, and monoclonal antibody A-I-57 that is the only one that gave parallel displacement curves between HDL and purified Apo A-I. As shown in **Fig. 3**, the displacement curves indicate that the binding of purified Apo A-I to lipids fully restores the immunoreactivity of the monoclonal antibodies to the protein and the binding affinity is almost the same for HDL and Apo A-I vesicles, thus suggesting that reversible conformational changes occurred in Apo A-I upon delipidization for which a particuliar epitope cannot be expressed.

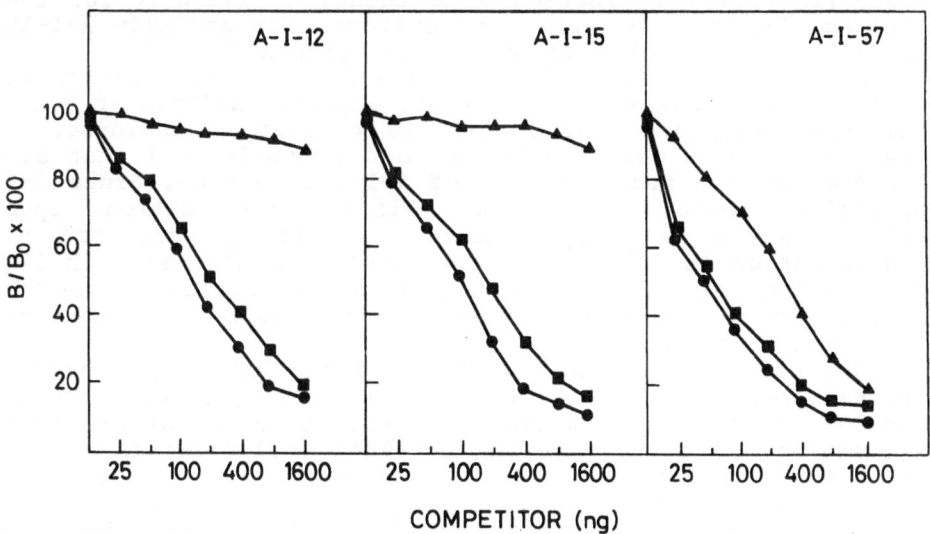

Figure 3. Displacement curves of ^{125}I-labeled HDL (●), purified apo A-I (▲), and apo A+I-egg+PC-vesicles (■). Apo A-I vesicles were prepared as described under Methods.

DISCUSSION

We described in this paper the effect of delipidation upon the presentation of epitopes for monoclonal antibodies raised against apolipoprotein A-I.

Our data clearly show that upon delipidization of Apo A-I the monoclonal antibodies   recognize less efficiently the apoprotein. This finding could be due to:

a) chemical modifications that occur during the delipidation and purification process

b) irreversible changes in the conformation of Apo A-I that do not allow proper presentation of the epitopes

c) reversible changes of the conformation that are restored upon reassociation of Apo A-I with lipids

d) self association of Apo A-I in the lipid free form that induces a partial or total shielding of epitopes.

All these possibilities may occur with apolipoprotein A-I. It has been shown, in fact, that chemical modification of Apo A-I occurs upon "aging" of the apolipoprotein (10) and that A-I undergoes formation of extensive alpha velical structure upon re-association with lipids (11). Finally self association of Apo A-I occurs in solution with formation of dimers, tetramers and octamers (12). All these observations may reflect changes in the presentation of epitopes.

Our experiments demonstrate that upon delipidization the epitopes for A-I-12 and A-I-15 are the most sensitive to the changes induces by the delipidization, A-I-57 is the antibody that most efficiently recognizes purified A-I in the free from, nevertheless also in this case a difference can be detected.

With the exception of A-I-12 and A-I-15 the slopes of the curves for HDL_3 or A-I are similar, but more purified A-I is required to obtain the effect. This is suggestive of a loss of the epitope in a certain percentage of Apo A-I molecules with no change in others resulting in much less immunoreactive A-I, instead of a change of affinity for the antigen. In the other hand the epitopes for antibody A-I-12 and A-I-15 are completely lost in purified A-I.

Upon reassociation of purified A-I with lipids the curve of competition comes back to almost superimpose that of HDL_3. This data suggests that reversible changes in conformation might occur upon delipidation of apolipoprotien A-I that are overcome by the reassociation of A-I with lipids. Since A-I partially   loses   its   alpha   helical   conformation   upon delipidization, and alpha helix formation is promoted by reassociation to lipids (11) we speculate that alpha helix distruption may be responsible, at least in part, for the loss of epitopes, that is partial for most antibodies or total for other antibodies.

One alternative explanation could be that a structure, not necessarely alpha helical, is restored by association of Apo A-I with lipids. During this process areas of A-I that are relatively distant in the unfolded (delipidated) state might become spatially close and contribute to the epitope for a MAB. Whether such a change can occur in all A-I molecules, and the factors controlling this phenomenon, is unknown.

In summary we have shown that   some   epitopes are expressed up to 100% in Apo A-I when present in HDL and may be partially lost upon delipidization of apolipoproteins. Our data suggest that changes in conformation of Apo A-I may contribute to this finding. Whether this phenomenon occurs in native lipoprotein when the lipoprotein structure is altered or the lipid composition is modified is unknown and remains to be tested as a possible mechanism of lipoprotein modification.

ACKNOWLEDGMENTS

This work was supported, in part, by grants from CNR. Miss Marta Colombo typed the manuscript.

REFERENCES

1) S. Eisenberg. High density lipoprotein metabolism. J. Lipid R. 25:1017 (1984).

2) S.W. Law, G. Gray, and H.B. Brewer, Jr. DNA cloning of human apo A-I: amino acid sequences of preproapoA-I. Biochem. Biophys. Res. Commun. 112:257 (1983).

3) §DC.J. Fielding, V.G. Shore, and P.E. Fielding. A protein cofactor of lecithin: cholesterol acyltransferase. Biochem. Biophys. Res. Commun. 46:1493 (1972).

4) G. Schmitz, H. Robenek, U. Lohmann, and G. Assman. Interaction of high density lipoproteins with cholesteryl ester-laden macrophages: Biochemical and morphological characterization of cell surface receptor binding, endocytosis and resecretion of high density lipoproteins by macrophages. EMBO J. 4:613 (1985).

5) S.H. Chem, C.Y. Yang, P.F. Ghen, D. Setzer, M. Tanimura, W-H. Li, A.M. Gotto Jr., and L. Chan. The complete cDNA and amino acid sequence of human apolipoprotein B-100. J. Biol. Chem. 261:12918 (1986).

6) R.J. Havel, H.A. Eder, and J.H. Bragdon. The distribution and chemical composition of ultracentrifugally separated lipoproteins in human serum. J. Clin. Invest. 34:1340 (1955).

7) C.E. Matz, and A. Jonas. Micellar complexes of human apolipoprotein A-I with phosphatidylcholinas and cholesterol prepared from cholate-lipid dispersions. J. Biol. Chem. 257:4535 (1982).

8) S. Marcovina, G. Di Cola, and A.L. Catapano Radial immunodiffusion assay of human apolipoprotein A-I with the use of two monoclonal antibodies combined. Clin. Chem. 32:2155 (1986).

9) D.W. Bilheimer S. Eisenbergh, and R.I. Levy. Metabolism of very low density lipoproteins Part 1 (Preliminary in vitro and in vivo observations). Biochilm. Biophys. Acta 260:212 (1972).

10) L.K. Curtiss, and R.S. Smith. Immunochemical heterogeneity of high density lipoproteins. Proceedings of the Workshop on Lipoprotein Heterogeneity. NIH Publication. 2646:366 (1987).

11) H.J. Pownall, Q. Pao, D. Hickson, J.T. Sparrow, S.K. Kosserow, and J.B. Massey. Kinetics and mechanism of association of human plasma apolipoproteins with dimyrisioyl phosphatidylcholine. Effect of protein structure and lipid clusters on reaction rates. Biochemistry. 20:6630 (1981).

12) Y.L. Marcel, D. Jewer, C. Vezina, P. Milthorp, and P.K. Weech. Expression of human apolipoprotien A-I epitopes in high density lipoproteins and in serum. J. Lipid Res. 28: 768 (1987).

FACTORS AFFECTING THE EXPRESSION OF APOLIPOPROTEIN A-I

EPITOPES AND SCREENING FOR MUTANTS

Y.L. Marcel, L. Leblond, E.A. Raffai, D. Jewer
P.K. Weech and R.W. Milne

Laboratory of Lipoprotein Metabolism
Clinical Research Institute of Montreal
110, Pine avenue west
Montreal, Quebec H2W 1R7, Canada

INTRODUCTION

Before applying monoclonal antibodies (Mabs) to the screening of apolipoprotein mutants, we should first consider the limitations of such an approach. To detect mutations randomly distributed along the primary sequence of an antigen, we need antibodies directed against epitopes that are also widely distributed. Early experimental evidence indicated that discrete antigenic epitopes exist, implying that specific regions of a protein are more antigenic than others. However as more experimental data accumulate, it appears that many overlapping epitopes exist on protein surfaces and that the whole surface may be antigenic[1]. Nevertheless, accessibility, flexibility and shape will determine which sites on proteins are most likely to be antigenic[2]. Novotny et al[3] have proposed that the primary reason for the antigenicity of certain polypeptide-chain segments is their exceptional surface exposure which make them readily available for contact with antibody combining sites. Because exposure of peptide segments results frequently in high mobility, the reported correlation between antigenicity and segmental flexibility may in fact be secondary to the first correlation between accessibility and antigenicity. These principles imply that, in the case of an apolipoprotein the aminoacids that constitute the hydrophobic face of the amphipathic α-helices and β-sheets and are in contact with the lipid phase of the lipoproteins, should be the least susceptible to represent the contact residues of an epitope. The opposite might be true for the amino acids constituting the hydrophilic face which should be both highly accessible and antigenic.

Identification of apolipoprotein A-I epitopes

In the case of apolipoprotein (apo)A-I, the molecule contains multiple repeats of 22 aminoacids, each of which is a repeat of 11-mers and which are sometimes arranged as 3 consecutive repeats. The repeat unit of 22-mer has been suggested to be a structural element that builds an amphipathic α-helix[4]. The mapping of apo A-I epitopes has not progressed sufficiently for us to know whether the α-helical 22-mers constitute antigenic domains. If as presumed these amphipathic repeats are bound to the lipid phase, they should be both poorly flexible and poorly accessible and thus of a limited antigenicity, whereas the β-turns which interupt the repeats should constitute more accessible and antigenic domains. One Mab has been obtained which reacts with an epitope having such a location: Ehnholm and colleagues[3] identified a Mab which reacts with normal apo A-I but not with a genetic variant of apo A-I ($Glu_{136} \rightarrow$ Lys). It has been proposed that this point mutation was the cause for the lack of reaction of this Mab with its epitope which is located between residues 113 and 148[5]. Residue 136 is indeed close to a proline at 139 which is thought to interupt the amphipathic helix[4]. Therefore it is possible that many of the helix breaking regions of apo A-I could constitute antigenic sites distributed over most of the apo A-I sequence. To date we have identified 2 epitopes in apo A-I CNBr-fragment 1, 1 epitope overlapping the CNBr 1-2 junction, 3 epitopes in fragment 2 and 2 epitopes in fragments 3[6-7]. Reports from other groups[5,8-9] have identified epitopes in different CNBr fragments including CNBr-fragment 4, but the number of distinct epitopes thus far identified on apo AI is uncertain. Accordingly and given the fact that apo AI epitope-mapping has been so far limited to their localization on CNBr fragments, the application of Mabs to the identification of apo A-I mutants can only be tentative for the moment.

Effect of lipids upon the expression of apo A-I epitope

Several studies have demonstrated that the well-documented heterogeneity of apo A-I containing lipoproteins resulted in the unequal and partial expression of many epitopes in the different lipoproteins. The first panel of Mabs described by Curtiss and Edgington[10] reacted with epitopes expressed on only 60% of the particles. While several of the Mabs from our laboratory reacted also with variable proportions of HDL ranging from 60 to 90%, 3 Mabs 2F2, 3G10 and 4F7 reacting on CNBr fragments 1, 2 and 3 respectively could precipitate 100% of HDL[11]. Another Mab binding to an epitope localized on CNBr fragment 1 was also reported to be expressed on all HDL particles[12]. It is thus certain that apo A-I epitopes are heterogeneously expressed on apo A-I containing lipoproteins, and that this heterogeneity is likely to be related at least in part to the heterogeneity of the apo A-I containing lipoproteins. Because the distribution of the different classes of apo A-I lipoproteins vary between subjects, this represents a limitation for the application of Mabs to the screening of apo A-I mutants. This should be best accomplished with Mabs reacting with well-characterized epitopes that are uniformly expressed independently of lipid composition and lipoprotein heterogeneity. Alternatively, Mabs reacting with partially expressed epitopes could also be used after a careful standardization of conditions with a large number of normolipemic plasma.

Amongst the Mabs characterized in our laboratory, 4H1 and 5F6 recognize antigenic determinants that are unaffected by the presence of lipids whereas antibodies 2F1, 5G6 and 4F7 do not react with isolated apo A-I but all of them are equally reactive with lipid-associated apo A-I independently of the unsaturation of the lipids. Finally other Mabs such as 3G10, 3D4 and 6B8 are partially reactive with delipidated apo A-I but more immunoreactive with apo AI associated with liposomes (unpublished results). This demonstrates that the expression of individual apo A-I epitopes may vary considerably depending on the presence of lipids and these properties must be taken into account when one evaluates the occurence of mutants in free apo A-I or lipoprotein-associated apo A-I.

The number of apo A-I molecules associated with plasma lipoproteins, which varies between particles represents an additional level of complexity in the immunochemistry of apo A-I. Nichols and associates[13,14] have shown that apo A-I containing lipoproteins may contain 2, 3 or 4 molecules of apo A-I per particle as a function of their sizes. Based on the hypothesis that nascent HDL particles are secreted by the liver or the intestine in the form of discoidal particles made up of cholesterol, lecithin, and apo A-I, these lipoprotein species are probably formed with a defined number of apo A-I molecules as a function of the relative amount of lipids and apolipoproteins that are available. Upon filling of the lipoprotein core with cholesteryl esters, species with 2 apo AI may fuse to generate species with 3 apo A-I while species containing 3 apo A-I may remain stable at that number[13,14]. We have hypothesized that owing to the different number of apolipoprotein molecules present in these lipoproteins, different packings of apo A-I protein chains may exist. This may in turn affect the conformation, flexibility and accessibility of different domains and hence the expression and immunoreactivity of specific epitopes. Indeed preliminary experiments have shown that the epitopes for Mabs 4H1 and 5F6 are clearly more immunoreactive in lipoproteins containing 2 apo A-I than in those containing 3 apo A-I whereas the epitopes for 3G10, 3D4 and 4F7 show little change in immunoreactivity (E. Raffai et al, unpublished results). There may be more than a simple coincidence in the fact that the antibodies 4H1 and 5F6, that react with epitopes independent of the presence of lipids, are influenced by the number of apo A-I per particle. We tentatively interpret these results as evidence that domains that are not involved in lipid binding may be more flexible and exposed and as such more susceptible to interaction with similar domains on other apolipoproteins. As a corollary, epitopes located on lipid binding domains are affected by the presence of lipids but these domains on the lipoprotein itself are more rigid and less susceptible to being influenced by the presence of other apolipoproteins. These examples underscore the complexity of apo A-I arrangement in lipoproteins and the resulting difficulty of its immunochemical evaluation.

Lability of apo A-I epitopes

We were the first to note the existence on apo A-I of epitopes for which immunoreactivity appeared to increase with time in vitro[15]. We identified 3 distinct epitopes with such properties which are distributed widely apart in the sequence indicating that this phenomenon was not related to a conformational modification at a single site. We have further shown that this effect is accelerated under alkaline conditions suggesting a possible role for deamidation[15]. Others[12] have made the same observations with a different panel of Mabs and identified other such labile epitopes that are presumably different from ours which demonstrates that this is a common phenomenon with Mabs specific for apo A-I. We have subsequently shown that the proportion of apo A-I containing lipoproteins expressing these epitopes increased with storage time in vitro and that the process was partially inhibited by EDTA and antioxidants[11]. Further work has now established that a peroxidative process is the mechanism by which certain apo A-I epitopes are developped and that the reaction requires the presence of polyunsaturated lipids (Marcel et al, manuscript in preparation).

In conclusion, the immunochemical characteristics of apo A-I, specially in the form of plasma lipoproteins, are very complex. The expression of apo A-I epitopes is often a function of the presence of lipids and of their nature, of the presence of other proteins, of the density of packing of apo A-I chains on the lipoproteins, and of in vitro and possibly in vivo deamidation and peroxidation reactions. Therefore the screening of populations for the occurence of apo A-I mutants should be done only with Mabs reacting with well-defined epitopes. While these requirements make the immunological screening of mutants a cumbersome approach, its advantage is the essential identification of mutations that result in significant changes in protein conformation and possibly in its metabolism.

Acknowledgements

The research of the authors cited here was supported by grants of the Medical Research Council of Canada (PG-27) and of the Quebec Heart Foundation. L.L. and E.A.R. were respectively fellow and medical scientist of the Canadian Heart Foundation. We acknowledge the expert editorial assistance of L. Lalonde.

REFERENCES

1.   D.C. Benjamin, J.A. Berzofsky, I.J. East, F.R.N. Gurd, C. Hannum, S.J. Leach, E. Margoiash, J.G. Michael, A. Miller, E.M. Prager, M. Reishlin, E.E. Sercarz, S.J. Smith-Gill, P.E. Todd, and A.C. Wilson. The antigenic structure of proteins: a reappraisal. Annu. Rev. Immunol. 2:67 ((1984).

2.   J. Novotny, M. Handschumaker and R.E. Bruccoleri. Protein antigenicity: a static surface property. Immunol. Today 8:26 (1987).

3.   J. Novotny, M. Handschumaker, E. Haber, R.E. Bruccoleri, W.B. Carlson, D.W. Fanning, J.A. Smith and G.D. Rose. Antigenic determinants in proteins coincide with surface regions accessible to large probes (antibody domaines). Proc. Natl. Acad. Sci. USA 83:226 (1986).

4.   W.-H. Li, M. Tanimura, C.-C. Luo, S. Datta and L. Chan. The apolipoprotein multigene family: biosynthesis, structure, structure-function relationships, and evolution. J. Lipid Res. 29:245 (1988).

5.   C. Ehnholm, M. Lukka, I. Rostedt and K. Harper. Monoclonal antibodies specific for different regions of human apolipoprotein A-I. Characterization of an antibody that does not bind to a genetic variant of apo A-I ($Glu_{1366}Lys$). J. Lipid Res. 27:1259 (1986).

6.   P.K. Weech, R.W. Milne, P. Milthorp and Y.L. Marcel. Apolipoprotein A-I from normal human plasma: definition of 3 distinct antigenic determinants. Biochim. Biophys. Acta 835:390 (1985).

7.   P. Milthorp, P.K. Weech, R.W. Milne and Y.L. Marcel. Immunological characterization of apolipoprotein A-I from normal human plasma. Mapping of antigenic determinants. In NATO ARW Proceedings vol. 112:103 Plenum Press N.Y. (1986).

8.   G. Schonfeld, E.S. Krul, R. Dargar and R.T. Kitchens. Epitope expression in HDL probed with monoclonal anti apo A-I antibodies. In Atherosclerosis VII: 255 Elsevier Science Published, Amsterdam (1986).

9.   E. Petit, M. Ayrault-Jarrier, D. Pastier, H. Robin, J. Polonovski, I. Aragon, E. Hervaud and B. Pau. Monoclonal antibodies to human apolipoprotein A-I. Characterization and application as structural probes for apo A-I and HDL. Biochim. Biophys. Acta 919:287 (1987).

10. L.K. Curtiss and T.S. Edgington. Immunochemical heterogeneity of human plasma high density lipoproteins. Identification with apolipoprotein A-I and A-II specific monoclonal antibodies. J. Biol. Chem. 260:2982 (1985).

11. Y.L. Marcel, D. Jewer, C. Vézina, P. Milthorp and P.K. Weech. Expression of human apolipoprotein A-I epitopes in high density lipoproteins and in serum. J. Lipid Res. 28:768 (1987).

12. L.K. Curtiss and R.S. Smith. Immunochemical heterogeneity of HDL in Proc. Workshop on Lipoprotein heterogeneity, p. 363, K. Lippel, Ed. NIH Publication No. 87-2646 (1987).

13. A.V. Nichols, P.J. Blanche, E.L. Gong, V.G. Shore and T.M. Forte. Molecular pathways in the transformation of model discoidal lipoprotein complexes induced by lecithin:cholesterol acyltransferase. Biochim. Biophys. Acta 834:285 (1985).

14. A.V. Nichols, E.G. Gong, P.J. Blanche, T.M. Forte and V.G. Shore. Pathways in the formation of human plasma high density lipoprotein subpopulations containing apolipoprotein A-I without apolipoprotein A-II. J. Lipid Res. 28:719 (1987).

15. P. Milthorp, P.K. Weech, R.W. Milne and Y.L. Marcel. Immunochemical characterization of apolipoprotein A-I from normal human plasma. In vitro modification of apo A-I antigens. Arteriosclerosis 6:285 (1983).

CHARACTERIZATION OF MONOCLONAL ANTIBODIES TO HUMAN LOW DENSITY LIPOPROTEIN

Alberto Corsini, Simona Fantappiè, Santica Marcovina*, Agnese Granata, Remo Fumagalli, and Alberico L. Catapano.

Institute of Pharmacological Sciences, and *S. Raffaele Hospital, University of Milan, Milan, Italy

Abstract : Nine monoclonal antibodies against human low density lipoprotein (LDL), that recognized apoprotein (apo) B-100, have been characterized in terms of specificity and ability to interfere with the receptor-mediated LDL catabolism in cultured human fibroblasts. Three of these antibodies inhibited the LDL-receptor interaction. The epitopes for these inhibitory antibodies have been assigned to the thrombolytic fragment T3 of apo B-100. These results are in agreement with other reports and suggest that the median portion of apo B-100 contains sequence(s) important in the LDL-receptor interaction.

INTRODUCTION

Epidemiological studies have shown a strong positive correlation between the levels of plasma LDL and the incidence of a coronary artery disease (1, 2). Plasma levels of LDL are determined by several processes, among them the receptor-mediated LDL catabolism plays a significant role (3-5). Apo B-100, a major protein constituent of LDL, is synthetized mainly by the liver, and is the ligand responsible for the interaction between LDL and their receptors (6). Because of the large size of apo B-100, its insolubility, and its propensity to break down into polypeptides, there was until recently little information on the structure and functional domains of this protein. Within the past three years the complete amino acid sequence of apo B-100 has been deduced from cDNA (7-10). Apo B-100 is a glycoprotein consisting of 4536 amino acids with a molecular weight of 520 kDa.

Inspection of the amino acid sequence of apo B suggests

that several regions have the potential to interact with the
ligand binding region of the LDL receptor (11). Studies with
monoclonal antibodies show that the area around the thrombin
cleavage site at apo B-100 residue 3249 (T2/T3) is
responsible for the interaction of apo B-100 with the LDL
receptor (7-12). We produced nine different monoclonal
antibodies against human LDL that recognize apo B-100. We
report on the specificity of these antibodies to apo B-100
and on their ability to interfere with the LDL binding to
the receptor in cultured human skin fibroblasts.

METHODS

Plasma lipoprotein isolation
LDL (d 1.019-1.063 g/ml) were isolated by sequential
preparative ultracentrifugation (13) and labeled with ^{125}I
according to Bilheimer et al. (14). Human lipoprotein-defi-
cient serum (LPDS) was prepared according to Brown et al.
(15). LDL apo B-100 was digested with thrombin (100:1, w:w)
at 25°C for 36 h in 10 mM Tris, 0.3 mM, pH 8.

Monoclonal antibodies
Monoclonal antibodies against LDL were prepared and characterized as described by Marcovina et al. (16).

Cells
Human skin fibroblasts, obtained from normolipidemic healthy
volunteers were grown in monolayers and maintained at 37°C
in a humified atmosphere of 95% air, 5% CO_2 in F-11 medium
supplemented with 10% fetal calf serum.

Experimental protocol
For all experiments, cells were seeded
in 35 mm dishes (1.5×10^5 cells) for 6 days. On day 6 the
medium was changed with one containing 10% LPDS and
confluent monolayers were incubated for 24h at 37°C. After
this time, fresh medium was added containing 7.5 ug of ^{125}
I-LDL protein/ml, and either 50 ug/ml of monoclonal antibodies or a 100 fold excess of unlabeled LDL, and the cells
were incubated for further 5h (this medium was preincubated
for 2 h at 37°C before being added to the cells). The uptake
and degradation of ^{125}I-LDL were determined as described by
Goldstein et al. (17).

Polyacrilamide gel electrophoresis and immunoblotting
Apo B and apo B-100 thrombolytic fragments were separated by
SDS polyacrylamide (4% and 6% respectively) gel electrophoresis (18) and transferred onto nitrocellulose according to
Towbin et al. (19). After saturation of nitrocellulose free
sites with PBS-BSA 3% at 37°C for 1 h, nitrocellulose strips
were incubated with monoclonal antibodies (100 ug protein in
5 ml PBS-BSA 0.5%) for 1 h at room temperature. Antibodies
were detected by incubation with a rabbit antibody against
mouse immunoglobulins labeled with ^{125}I by the Iodogen
procedure (20).

RESULTS AND DISCUSSION

To localize the region(s) of apo B-100 involved in the recognition of the LDL receptor, we tested all the monoclonal antibodies (MABS) used in the present study for their ability to inhibit the uptake and degradation of ^{125}I-LDL to the LDL receptor by cultured fibroblasts.

As shown in the **Table1**, only three monoclonal antibodies interferred with LDL-receptor interaction. Specific uptake and degradation of ^{125}I-LDL were inhibited up to 75% by antibodies 2A, 7A and 9A.

TABLE 1

EFFECT OF MONOCLONAL ANTIBODIES TO APO B-100 ON THE UPTAKE AND DEGRADATION OF ^{125}I-LDL BY HUMAN SKIN FIBROBLASTS

| | ANTIBODY (50 ug/ml medium) | | | | | | | | | |
	none	2A	3B	3H	4B	6B	6E	7A	8A	9A
UPTAKE	82.7	27.3	92.6	111.8	75.2	101.0	89.1	22.6	79.2	21.7
DEGRADATION	157.1	47.6	190.1	212.9	146.2	167.8	146.9	43.0	132.8	38.1

Experimental conditions as described in Methods; data are the mean of triplicates that did not differ by more than 10% and are expressed as ng of ^{125}I-LDL protein/mg cell protein.

The epitopes of these antibodies have been mapped to the apo B-100 thrombolytic fragments (T1, 385 kDa; T2, 170 kDa; T3, 283 kDa; T4, 145 kDa), where T3 and T4 were derived from further digestion of T1 (21). As depicted in the **Figure1**, the determinants for seven monoclonal antibodies were located on fragment T3; one (MAB 3H) was found in T2 and another (6E) in T4. The assignement of the epitopes for all the inhibitory antibodies 2A, 7A, 9A to the median fragment T3 of apo B-100, provides evidence for the role of this region in receptor binding and corrborates the earlier identification of potential receptor-binding sequences clustered around the T3/T2 junction (7, 12).

237

We do not know whether the ability of these three antibodies to inhibit LDL-receptor binding reflects the proximity of their epitopes to the receptor binding domain(s). These antibodies may interfere with the orientation of the LDL particle on the cell surface and thus hinder the interaction of apo B-100 with the LDL receptor. Alternatively, monoclonal antibodies may induce a conformational change in apo B which decreases the affinity of LDL for the receptor. Finally, the epitopes of these inhibitory antibodies may be distant from the receptor binding sequences in the primary structure of apo B but relatively close in the native LDL particle due to the protein folding. At present we have no data to prove which hypothesis is correct.

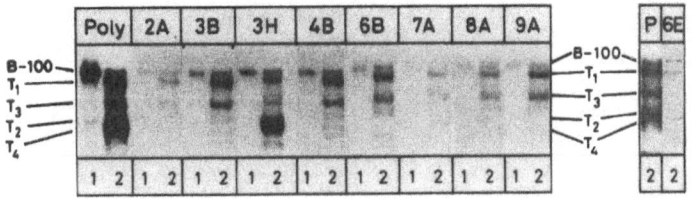

Figure1 Recognition of apolipoprotein B-100 thrombolytic fragments by murine monoclonal antibodies to LDL. Lane 1: apo B-100; lane 2: thrombolytic fragments of apo B-100; poly: polyclonal antibody.

In summary, consistent with previous reports (7, 12), our results suggest that a median portion of apo B-100, represented by T3, is important in determining the LDL-receptor interaction.

Studies are in progress to investigate in more details the epitopes location for our inhibitory monoclonal antibodies to gain further insights on the localization and nature of the receptor-binding domain(s) of apo B-100.

ACKNOWEDGMENTS

This work was supported in part by a CNR grant (Progetto Finalizzato Malattie Degenerative) to ALC and SM, and by MPI, Italian Government. Miss Marta Colombo typed the manuscript.

REFERENCES

1) J. Stamler, Population studies, in: Nutrition, Lipids, and Coronary Heart Disease. R. Levy, B. Rifkind, B. Dennis, and N. Ernst, editors. Raven Press, New York. 25 (1979).

2) Lowering Blood Cholesterol to Prevent Heart Disease, Consensus Conference. J. Am. Med. Assoc. 253:2080 (1985).

3) R.J. Havel, J.L. Goldstein, and M.S. Brown. Lipoproteins and lipid transport, in: Metabolic Control and Disease. 8th Edition. P.K. Bondy, and L.E. Rosenberg, editors. W.B. Saunders, Philadelphia, PA. 393 (1980).

4) J.P. Kane, Apolipoprotein B: structural and metabolic heterogeneity. Annu. Rev. Physiol. 45:637 (1983).

5) J.L. Goldstein, and M.S. Brown. Familial hypercholeste- rolemia, in: The Metabolic Basis of Inherited Disease. 5th Edition. J.B. Stanbury, J.B. Wyngaarden, D.S. Fredrickson, J.L. Goldstein, and M.S. Brown, editors. McGraw-Hill, New York 672 (1983).

6) M.S. Brown, and J.L. Goldstein. A receptor-mediated pathway for cholesterol homeostasis. Science. 232:34 (1986).

7) T.J. Knott, R.J. Pease, L.M. Powell, S.C. Wallis, S.C. Rall, Jr., T.L. Innerarity, B. Blackhart, W.H. Taylor, Y. Marcel, R. Milne, D. Johnson, M. Fuller, A.J. Lusis, B.J. McCarthy, R.W. Mahley, B. Levy-Wilson, and J. Scott. Complete protein sequence and identification of structural domains of human apolipoprotein B. Nature. 323:734 (1986).

8) S-H. Chen, C-Y. Yang, P-F. Chen, D. Setzer, M. Tanimura, W-H. Li, A.M. Gotto, Jr., and L. Chan. The complete cDNA and amino acid sequence of human apolipoprotein B-100. J. Biol. Chem. 261:12918 (1986).

9) S.W. Law, S.M. Grant, K. Higuchi, A. Hospattankar, K. Lackner, N. Lee, and H.B. Brewer, Jr. Human liver apolipo- perotein B-100 cDNA: complete nucleic acid and derived amino acid sequence. Proc. Natl. Acad. Sci. USA. 83:8142 (1986).

10) C. Cladaras, M. Hadzopoulou-Cladaras, R.T. Nolte, D. Atkinson, and V.I.Zannis. The complete sequence and structu- ral analysis of human apolipoprotein B-100: relationship between apo B-100 and apo B-48 forms. EMBO (Eur. Mol. Biol. Organ.) J. 5:3495 (1986).

11) A.V. Hospattankar, S.W. Law, K. Lackner, and H.B. Brewer Jr. Identification of low density lipoprotein receptor binding domains of human apolipoprotien B-100: a proposed consensus LDL receptor binding sequence of Apo B-100. Biochem. Biophys. Res. Commun. 139:1078 (1986).

12) Y.L. Marcel, T.L. Innerarity, C. Spilman, R.W. Mahley, A.A. Protter, and R.W. Milne. Mapping of human apolipo- protein B antigenic determinants. Arteriosclerosis. 7:166 (1987).

13) R.J. Havel, H.A. Eder, and J.H. Bragdon. The distribu- tion and chemical composition of ultracentrifugally separa- ted lipoproteins in human serum. J. Clin. Invest. 34:1345 (1955).

14) D.W. Bilheimer, S. Eisenberg, and R.I. Levy. The metabolism of very low density lipoprotein proteins. I. Preliminary in vitro and in vivo observations. Biochim. Biophys. Acta. 260:212 (1972).

15) M.S. Brown, S.E. Dana, and J.L. Goldstein. Regulation of 3-hydroxy-3-methylglutaryl coenzime A reductase activity in cultured human fibroblasts. J. Biol. Chem. 249:789 (1974).

16) S. Marcovina, D. France, R.A. Phillips, and S.J.T. Mao. Monoclonal antibodies can precipitate low-density-lipoprotein. Characterization and use in determining apolipoprotein B. Clin. Chem. 31:1654 (1985).

17) J.L. Goldstein, S.K. Basu, and M.S. Brown. Receptor-mediated endocytosis of low density lipoproteins in cultured cells. Methods Enzymol. 98:241 (1983).

18) U.K. Laemmli. Cleavage of structural proteins during the assembly of the head of bacteriophage T4. Nature (London). 227:680 (1970).

19) H. Towbin, T. Staehelin, and J. Gordon. Electrophoretic transfer of proteins from polyacrylamide gels to nitrocellulose sheets: procedure and some applications. Proc. Natl. Acad. Sci. USA. 76:4350 (1979).

20) P.J. Fraker, and J.C. Speck, Jr. Protein and cell membrane iodinations with a sparingly soluble chloroamide 1, 3, 4, 6-tetrachloro 3a, 6a diphenylglycoluril. Biochem, Biophys, Res. Commun. 80:849 (1980).

21) A.D. Cardin, K.R. Witt, J. Chao, H.S. Margolius, V.H. Donaldson, and R.L. Jackson. Degradation of apolipoprotein B-100 of human plasma low density lipoproteins by tissue and plasma kallikreins. J. Biol. Chem. 259:8522 (1984).

APOLIPOPROTEIN B : IMMUNOLOGICAL METHODS FOR THE DETECTION OF MUTANTS

P. Duriez[1,2], A. Dunning[3], N. Vu Dac[1], F. Monard-Herkt[1],
H. Parra[1], S. Humphries[3], J.C. Fruchart[1]

1 - SERLIA, Institut Pasteur et Inserm U. 279, 1 rue du
    Professeur Calmette,59019 Lille Cédex
2 - Laboratoire de Physiologie, Faculté de Médecine,Lille
3 - Charing Cross Sunley Research Centre, London

The conventional approach to the study of dyslipoproteinemia has been to look for abnormalities in protein structure and function in the different defined disorders. This has led to identification of genetic defects in the cell surface receptor that control the degradation of LDL[1]. Furthermore, a mutant form of apolipoprotein A-I (apo A-I Milano)[2] associated with hypertriglyceridemia and variants of apolipoprotein E[3] associated with type III hyperlipoproteinemia have been demonstrated in the population. Recently, 3 polymorphic forms of apo A-I have been identified by isoelectric focusing gels[4].

Although these different polymorphic species of apolipoproteins vary in size, in charge and in amino acid sequence, their identification by immunological means has not yet been documented.

The Ag system of lipoproteins

The first immunologically detected polymorphism of human serum lipoproteins was reported by Allison and Blumberg[5] ; it was later shown to relate to LDL and designated the Ag system. This system comprises 5 pairs of antigens (a1,d ; c,g ; t,z ; h,i ; x,y) behaving as products of allelic genes[6]. To date, no common immunogenically revealed polymorphisms of HDL have been described in any species other than the rabbit[7].

The existence of genetically determined variants of LDL is of interest in light of the role which this lipoprotein plays in atherosclerosis. The plasma levels of LDL are determined in large part by LDL receptors which recognize and bind the protein moiety of LDL, apolipoprotein B_{100}[1]. The importance of LDL receptors in the regulation of plasma LDL concentration is illustrated by the genetic disorder familial hypercholesterolemia. Affected individuals either lack or have defective receptors and develop hypercholesterolemia because of impaired plasma clearance of LDL[1]. However, most individuals with elevated cholesterol levels apparently possess normal LDL receptors. Thus, in most cases of primary hypercholesterolemia, other abnormalities must be responsable for elevated LDL levels. In theory variations in the major LDL apoprotein, apo B_{100}, could lead to defects in LDL catabolism since this protein plays a key role in the interaction between LDL and its receptor. Recently, the existence of hypercholesterolemic patients with normal LDL receptors but abnormal LDL has been reported[8]. These LDL exhibit defective receptor binding resulting in inefficient clearance of LDL and the hypercholesterolemia observed in these patients.

In pigs, genetic polymorphism of LDL has been strongly correlated with premature atherosclerosis[9]. However, in man the relevance of Ag variations to atherosclerosis has not been firmly established and may be limited to the effect of a component of this system termed Ag(x) on lipid levels[10].

Ag antisera, obtained from multiply transfused patients, are always in short supply because only a small number of heavily transfused individuals produce them[11]. This difficulty in obtaining antisera is partly responsible for the scant attention that has been paid to the Ag system in atherosclerosis research. Preliminary experiments to obtain Ag antisera in animals have not succeeded[11]. However, recourse to monoclonal antibodies has given us an opportunity to obtain large quantities of well characterized Ag typing reagents despite such problems.

Ag system recognition by monoclonal antibodies

Recently, two murine monoclonal antibodies (MB19[12-16] and BIP 45[17]) to an Ag determinant (the Ag(c) antigen) have been produced.

Three types of evidence suggest that these monoclonal antibodies recognize the Ag(c) factor. First, individuals homozygous for the Ag(c) allele are strong reactors against MB19 and BIP 45 while those homozygous for the Ag(g) allele are weak reactors against these monoclonal antibodies[13-17]. Moreover, individuals heterozygous for the Ag alleles c and g are intermediate reactors (Table 1). Secondly, in family studies, the antigens recognized by MB19[13-16] and BIP 45[17] and their homologues, that are not recognized by these monoclonal antibodies, behave as products of 2 allelic genes which are expressed in an autosomal codominant fashion (Fig. 1). This genetic transmission corresponds to the mode of transmission of the c and g factors[6]. Thirdly, the gene frequency for alleles detected by MB19[13-16] and by BIP 45[17] antibodies corresponds to the gene frequency at the Ag(c)/Ag(g) locus in a caucasian population[6] (Table 2).

Since both monoclonal antibodies recognize the Ag(c) epitope, it was of interest to determine if they compete for LDL binding. In such competitive binding studies it was shown that ^{125}I-labeled BIP 45 was specifically and effectively displaced by MB19 from binding to LDL[18]. This result indicates that MB19 and BIP 45 recognize the same or closely related epitopes on LDL.

Thus, use of monoclonal antibodies against the Ag(c) antigen of LDL has given information on (a) the biochemical nature of the Ag polymorphism, (b) the genetic origin of apolipoprotein B_{100}, apo B_{48} and apo B_{37}, (c) the influence of the Ag(c,g) system on the apo B binding to its receptor and (d) the clinical relevance of the Ag (c,g) polymorphism.

a) The biochemical nature of the Ag polymorphism. MB19 was shown to bind to the protein portion of apolipoprotein B, since removal of carbohydrate from the LDL does not after binding[12] and it detects apolipoprotein B after SDS gel electrophoresis and transfer to nitrocellulose (Western blots)[19]. Since all the Ag loci are closely linked, this evidence strongly suggested that the Ag system is located on the structural gene for apolipoprotein B[14].

This hypothesis has recently been confirmed by the use of RFLP's at the apo B locus ; the XbaI RFLP of the apo B gene has been shown to be closely linked to the Ag(x,y)[20] and the Ag(c,g)[21-23] systems of immuno-

Table 1. Expression of Ag phenotypes and of apo B reactivity with monoclonal BIP 45 antibody in unrelated individuals

S = strong reactor, I = intermediate reactor, W = weak reactor,
C = concentration of apo B as competitor when $B/B_0 = 0.25$,
where B corresponds to the absorbance in the presence and B_0
in the absence of competing apo B.

n	x	y	al	d	t	z	h	i	c	g	Apo B binding to BIP 45 $C(\mu g/ml)$	
0	+	−	+	−	+	−	−	+	−	+	95	W
8	+	+	+	−	+	+	−	+	−	+	60	W
11	+	+	−	+	+	−	−	+	−	+	75	W
13	−	+	+	−	+	+	−	+	−	+	75	W
15	+	+	+	−	+	+	−	+	−	+	65	W
16	+	+	+	−	+	+	−	+	−	+	200	W
21	−	+	−	+	+	−	−	+	−	+	70	W
24	−	+	+	−	−	+	−	+	−	+	150	W
26	−	+	+	−	+	+	−	+	−	+	55	W
31	+	+	+	+	+	−	−	+	−	+	115	W
33	+	−	+	−	+	−	−	+	−	+	70	W
53	+	+	+	−	+	−	−	+	−	+	90	W
72	+	+	+	−	+	−	−	+	−	+	60	W
108	+	−	+	−	+	−	−	+	−	+	105	W
168	+	−	+	−	+	−	−	+	−	+	70	W
T168	+	−	+	−	+	−	−	+	−	+	68	W
209	+	−	+	+	+	−	−	+	−	+	75	W
5	−	+	−	+	+	−	−	+	+	+	25	I
10	−	+	+	+	+	−	+	+	+	+	26	I
14	−	+	−	+	+	−	−	+	+	+	30	I
30	+	+	+	+	+	+	−	+	+	+	21	I
32	−	+	−	+	+	−	+	+	+	+	35	I
35	+	+	+	+	+	−	−	+	+	+	25	I
36	−	+	+	+	+	+	+	+	+	+	22	I
42	+	+	+	+	+	−	+	+	+	+	29	I
20	−	+	−	+	+	−	+	−	+	−	13	S
39	−	+	−	+	+	−	−	+	+	−	15	S
91	−	+	+	+	+	+	−	+	+	−	13	S
405	−	+	−	+	+	−	+	−	+	−	17	S
408	−	+	−	+	+	−	+	−	+	−	11	S

S : strong reactor
I : intermediate reactor
W : weak reactor against
BIP 45

Table 2. Determination of the theoretical distribution of the reference population into 3 genotypes according to Hardy Weinberg statistics using LDL binding affinity against BIP 45.

Genotype	Experimental reference population n = 244	Allele frequencies	Theoretical distribution Hardy Weinberg equilibrium ($X^2 = 0.487$ NS)
BIP⁻/BIP⁻ Ag(c^-,g^+)	114 (46.7%)		$p^2 = 47.6\%$
		Ag(g): p = 0.69	
BIP⁻/BIP⁺ Ag(c^+,g^+)	109 (44.7%)		$2pq = 42.8\%$
		Ag(c): q = 0.31	
BIP⁺/BIP⁻ Ag(c^+,g^-)	71 (8.6%)		$q^2 = 9.6\%$

genetic variants while the EcoRI RFLP have been shown to be very strongly associated with the Ag(t,z) immunochemical polymorphism of human LDL[21].

The XbaI mutation is a silent C to T mutation. That is, it does not result in an amino acid alteration. Therefore, it cannot coincide with the Ag(c,g) locus, a result already anticipated from the imperfect association betwteen Ag(c,g) allotypes and XbaI RFLP[21-23]. The XbaI mutation is located at residue 2488, while the Ag(c,g) mutation is located on the N-terminal thrombolytic peptide T4 of apo B, which comprises residues 1 to 1297 on the mature protein[14]. Thus, the two sites would appear to be separated by at least 1191 residues, about one-quarter of the length of apo B. Associations between the epitopes detected by MB19[22] and BIP 45[23] (Table 3) monoclonal antibodies and these XbaI polymorphisms have been found.

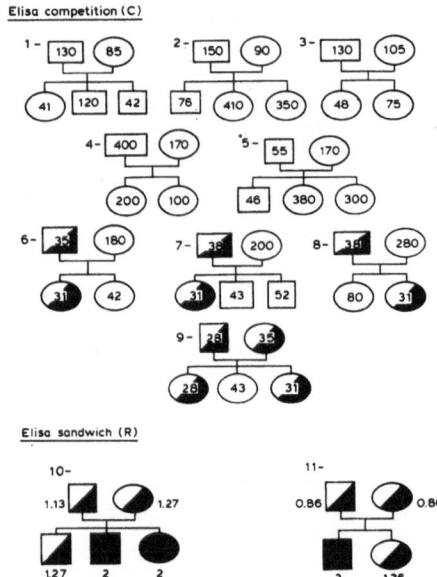

Fig. 1 Family studies. Females are shown by circles, males are shown by squares. Homozygotes for the apo B allotype with high affinity for BIP 45 are shown by a black square or circle. Individuals homozygous for the apo B allotype with low affinity for BIP 45 are shown by a white square or circle and those heterozygous for apo B allotypes are expressed by a white-and-black square or circle.

The EcoRI site polymorphism alters the apo B protein sequence by substituting lys for glu at residue 4154 in the mature protein[20] and is a candidate for being the structural alteration recognized by the Ag(t,z) antibodies, although this remains to be proven[21]. In any case, the RFLP closely associated with the Ag(t,z) epitope is located toward the c-terminus of apo B[21].

These data are consistent with the concept that the Ag(x,y), the Ag(c,g) and the Ag(t,z) antigens are true genetic polymorphisms of apo B ; that is the variations are not due to differences in post-translational processing of the protein or in its lipid binding capacity.

Table 3. Association between BIP 45 binding affinity and XbaI RFLP.

		BIP-45 Binding Affinity			
		gg Low	cg Med	cc High	
Pvu II genotype	P1P1	21	16	6	$X^2 = 1.14$ N.S.
	P1P2	5	4	1	$\Delta = 0.09$ confid. $= 0.28$ & limits -0.1
	P2P2	1	0	0	Frequency P1 $= 0.89$ P2 $= 0.11$
EcoRI genotype	R1R1	14	7	5	$X^2 = 5.93$ N.S.
	R1R2	12	10	1	$\Delta = 0.04$ confid. $= 0.23$ & limits -0.15
	R2R2	1	3	0	Frequency R1 $= 0.71$ R2 $= 0.29$
Xba I genotype	X1X1	11	3	0	$X^2 = 10.51$ $P < 0.05$
	X1X2	9	13	4	$\Delta = 0.32$ confid. $= 0.48$ & limits 0.14
	X2X2	7	3	4	Frequency X1 $= 0.50$ X2 $= 0.50$

Frequency g $= 0.69$
c $= 0.31$

b) The genetic origin of apolipoprotein B_{100}, apo B_{48} and apo B_{37}.
Immunological studies have shown that the MB19 polymorphism is present
both in apo B_{100} and apo B_{48}[19]. This observation provides evidence that
both apolipoproteins are products of the same gene and is in agreement
with recent molecular analysis of the apo B gene[24,25].

Recently, a kindred with familial hypobetalipoproteinemia of 41
family members in three generations has been studied[26,27]. The presence
of 2 distinct apo B alleles associated with low plasma concentrations of
apo B and LDL cholesterol and the inheritance of these 2 alleles have
been described. One of the alleles resulted in the production of an
abnormal, truncated apo B species, apo B_{37} and of abnormal lipoprotein
particles. The other apo B allele was associated with reduced plasma
concentrations of the normal apo B species, apo B_{100}. In this family,
MB19 bound very strongly to apo B_{37}, indicating that this protein was
the product of an allele encoding for allotype Ag(c) whereas it bound
very weakly to the apo B_{100} species and was associated with reduced
concentrations of apo B, indicating that this protein was the product of
an apo B allele encoding for allotype Ag(g). So, the immunochemical
genetic polymorphism in apo B can be used as a tool to identify abnormal
apo B alleles.

**c) The influence of the Ag(c,g) system on apo B binding to its
receptor.** Intestinal apo B_{48} represents the amino-terminal 2152 amino
acids of apo B_{100}[25] and apo B_{37} contains amino-terminal domains of apo
B_{100}[26]. The allelic variant recognized by MB19 has been shown to present
on apo B_{100}, apo B_{48} and apo B_{37}. Thus, the epitope corresponding to the
Ag(c) locus must be located on a region of amino-acid sequence common to
apo B_{100}, apo B_{48} and apo B_{37}. On the other hand the apo B receptor
binding domain is localized on the c-terminus of apo B_{100}[28,29]. Apo

B_{48}[30] and apo B_{37}[31] lipoprotein particles do not bind to the LDL receptor, therefore, it might be supposed that the Ag(c,g) locus would not influence the binding of LDL to its receptor. This has been studied with cultured fibroblasts where it has been shown that when bound to LDL, MB19 did not inhibit the uptake and degradation of LDL by the LDL receptor[16,32]. Not surprinsingly then the Ag(c,g) polymorphism is not associated with large variations in apo B and LDL cholesterol levels. This has been confirmed in clinical studies.

d) The clinical relevance of Ag(c,g) polymorphism. Studies in large populations with MB19[16,32] and BIP 45[33] have shown that the Ag (c,g) polymorphism of apo B is not associated with variations in lipoprotein levels in the different populations. However, average apo B and/or cholesterol concentrations are slightly (but not with significantly) higher in the groups having an intermediate binding affinity for MB19 and BIP 45 (Table 4) monoclonal antibodies (genotype Ag(c^+,g^+)) versus those with low (genotype Ag(c^-,g^+)) binding affinity. Furthermore, the lowest apo B and/or cholesterol levels were in groups with the higher affinity binding (genotype Ag(c^+,g^-)). It is remarkable that 3 different clinical studies show the same pattern. Moreover, when we studied the relationships between the epitope recognized by BIP 45 and RFLPs of the gene for apo B detected by XbaI[23] we have found associations between binding affinity to BIP 45 and one XbaI RFLP and between this RFLP and serum cholesterol levels (Table 5) but we have not found any significant direct association between serum cholesterol levels and BIP 45 binding affinity (total cholesterol was slightly increased, without signifi- cance, in the Ag(c^+,g^+) group). Since the MB19 monoclonal antibody does not inhibit the uptake and the degradation of LDL by cultured fibro- blasts[16,32] it is possible that the increased level of apo B and/or cholesterol in the Ag(c^+,g^+) group may be caused by a rise in apo B synthesis rather than a defect in its catabolism. This hypothesis needs further investigation.

Clinical studies have shown that the population frequencies of the 3 polymorphic groups defined by MB19[16,32] and BIP 45[33] were not dif- ferent in familial hypercholesterolemic subjects compared to reference populations. These data support the conclusion that the Ag(c,g) polymor- phism is not associated with the presence of familial hypercholestero- lemia. However, since the gene coding for the LDL receptor and that coding for apo B are located on different chromosomes[1,24] there is no reason to expect a cosegregation between familial hypercholesterolemia and apo B type.

Table 4. Plasma lipids and apolipoprotein B in the 3 subgroups of the reference population (Ag(c^-,g^+) ; Ag(c^+,g^+) ; Ag(c^+,g^-)).

Phenotype	n	cholesterol (mmol/l)	triglyceride (mmol/l)	apo B (g/l)
Ag(c^-,g^+)	114	6.14 + 1.12	1.47 + 0.97	0.95 + 0.37
Ag(c^+,g^+)	109	6.28 + 1.13	1.31 + 0.66	1.02 + 0.40
Ag(c^+,g^-)	21	5.98 + 0.93	1.14 + 0.57	0.86 + 0.28

Mean + SD
Non parametric statitics (Mann- Whitney test)

		NS	NS	NS

Table 5. Effect of XbaI genotype and BIP 45 binding-affinity on serum
cholesterol level.

			Number	Mean serum cholesterol	Std. dev.	
XbaI	1	1	14	6.83	0.70	F = 4.10
genotype	1	2	25	6.32	1.04	df 2 48
	2	2	12	5.78	0.94	p = 0.025
BIP 45 binding affinity	Low g g		28	6.28	0.89	F = 1.65
	Med c g		17	6.62	1.01	df 2 49
	High c c		7	5.85	1.19	p = 0.25

A cardiological study published using MB19[32] has indicated that the
Ag(c,g) polymorphism is not linked to coronary artery disease. The study
performed with BIP 45[33] suggested that there might be an increase in the
frequency of the Ag(c$^+$,g$^+$) phenotype in patients suffering from coronary
artery disease ; on the other hand, the frequency of the Ag(c$^-$,g$^+$)
phenotype was decreased in this population (Table 6). However, analysis
of the variance of the allele Ag(g) indicated that this difference may
not be significant. Nevertheless, the Ag(c,g) phenotypes cannot be
excluded from having a functional role in the development of atherosclerosis. Studies are presently under way to find and further investigate these functional variants of apolipoprotein B.

CONCLUSION

These results indicate this the use of monoclonal antibodies may be
helpful to study the clinical relevance of expressed polymorphisms in
apolipoproteins. It has been demonstrated that monoclonal or oligoclonal
antibodies may be prepared against synthetic peptides of different
portions of apolipoproteins[34]. In the future, it may be expected that
such antibodies directed against synthetic peptides of defined mutant
portions of apolipoproteins could be of substantial clinical relevance.
These antibodies together with methods suitable for mass screening such
as ELISA would facilitate the monitoring of apolipoprotein polymorphisms
in large populations.

Table 6. Comparison of the distribution of the reference
population (REF) with the population without
(CAD$^-$) and with (CAD$^+$) angiographically proven
coronary artery disease.

Phenotype	Ref n = 244	CAD- n = 24	CAD+ n = 64
Ag(c-,g+)	46.7%	45.5%	25%
Ag(c+,g+)	44.7%	45.5%	64%
Ag(c+,g-)	8.6%	9%	11%
		x^2 = 0.047	x^2 = 9.91
		NS	p < 0.01

REFERENCES

1. M. S. Brown and J. L. Goldstein, A receptor-mediated pathway for cholesterol homeostasis, Science 232:34 (1986).

2. K. H. Weisgraber, T. P. Bersot, R. W. Malhey, G. Franceschini and C. R. Sirtori, A-I Milano apoprotein. Isolation and characterization of a cysteine containing variant of the A-I apoprotein from human high density lipoproteins, J. Clin. Invest. 66:901 (1980).

3. K. H. Weisgraber, S. C. Rall, Jr and R. W. Malhey, Human E apoprotein heterogeneity cysteine arginine interchanges in the amino acid sequence of the apo E isoforms, J. Biol. Chem. 256:9077 (1981).

4. H. J. Menzel, G. A. Smann, S. C. Rall, K. H. Weisgraber and R. Malhey, Human apolipoprotein A-I polymorphism. Identification of amino acid substitutions in three electrophoretic variants of the Münster 3 type, J. Biol. Chem. 259:3070 (1984).

5. A. C. Allison and B. S. Blumberg, An isoprecipitation reaction distinguishing human serum protein types, Lancet 1:634 (1961).

6. R. Butler, E. Brunner, and G. Morganti, Contribution to the inheritance of Ag-groups ; a population genetic study, Vox Sang 26:485 (1974).

7. K. Berg, Genetics of coronary heart disease, in: "Progress in Medical Genetics," A. G. Steinberg, A. G. Bearn, A. G. Motulski and B. Childs, eds., Saunders, Philadelphia (1983).

8. T. L. Innerarity, K. H. Weisgraber, K. S. Arnold, R. W. Mahley, R. M. Krauss, G. L. Vega and S. M. Grundy, Familial defective apolipoprotein B_{100} : low density lipoproteins with abnormal receptor binding, Proc. Nat. Acad. Sci. USA 84:6919 (1987).

9. J. Rapacz, J. Hasler-Rapacz, K. M. Taylor, W. J. Checouich and A. D. Attie, Lipoprotein mutations in pigs are associated with elevated plasma cholesterol and atherosclerosis, Science 234:1573 (1986).

10. K. Berg, C. Hames, G. Dahlen, M. H. Frick and I. Krisham, Genetic variation in serum low density lipoproteins and lipid levels in man, Proc. Nat. Acad. Sci. USA 73:937 (1976).

11. K. Berg, Genetic variation in low density lipoprotein. Beginning of a new era ? J. Immunogenetics 12:263 (1986).

12. V. N. Schumaker, M. T. Robinson, L. K. Curtiss, R. Butler and R. S. Sparkes, Anti-apoprotein B monoclonal antibodies detect human low density lipoprotein polymorphism, J. Biol. Chem. 259:6423 (1984).

13. S. G. Young, S. J. Bertico, L. K. Curtiss, D. C. Casal and J. L. Witztum, Monoclonal antibody MB 19 detects genetic polymorphism in human apolipoprotein B, Proc. Nat. Acad. Sci. USA 83:1101 (1986).

14. M. T. Robinson, V. N. Shumaker, R. Butler, K. Berg and L. K. Curtiss, Ag(c) : recognition by a monoclonal antibody, Arteriosclerosis 6:341 (1986).

15. M. J. Tikkanen, M. J. Ehnholm, R. Butler, S. G. Young, L. K. Curtiss and J. L. Witztum, Monoclonal antibody detects Ag polymorphism of apolipoprotein B, FEBS Lett 202:54 (1986).

16. M. J. Tikkanen, C. Ehnholm, P. T. Kovanen, R. Butler, S. G. Young, L. K. Curtiss and J. L. Witztum, Detection of two apolipoprotein B species (apo Bc and apo Bg) by a monoclonal antibody, Atherosclerosis 65:247 (1987).

17. P. Duriez, R. Butler, M. J. Tikkanen, J. Steinmetz, N. Vu Dac, E. Butler-Brunner, I. Luyeye, J. M. Bard, P. Puchois and J. C. Fruchart, A monoclonal antibody (BIP 45) detects Ag(c,g) polymorphism of human apolipoprotein B, J. Immunol. Meth. 102:205 (1987).

18. M. J. Tikkanen, unpublished results (1987).

19. S. G. Young, S. J. Bertics, T. M. Scott, B. W. Dubois, L. K. Curtiss and J. L. Witztum, Parallel expression of the MB19 genetic polymorphism in apoprotein B_{100} and apo B_{48}, J. Biol. Chem. 261:2995 (1986).

20. K. Berg, L. M. Powell, S. C. Wallis, R. Pease, T. J. Knott and J. Scott, Genetic linkage between antigenic group (Ag) variation and the apolipoprotein B gene : assignment of the Ag locus, Proc. Nat. Acad. Sci. USA 83:7367 (1986).

21. Y. Ma, V. N. Shumaker, R. Butler and R. S. Sparkes, Two DNA restriction fragment length polymorphism associated with Ag (t/z) and Ag(g,c) antigenic sites of human apolipoprotein B, Arteriosclerosis 7:301 (1987).

22. A. M. Dunning, M. J. Tikkanen, C. Ehnholm, R. Butler and S. E. Humphries, Relationships between DNA and protein polymorphisms of apolipoprotein B, Hum. Gen. (in press) (1988).

23. A. Dunning, P. Duriez, N. Vu Dac, J. C. Fruchart and S. E. Humphries, Association between epitopes detected by monoclonal antibody BIP 45 and the XbaI polymorphism of apolipoprotein B, Clin. Genet. 33:181 (1988).

24. K. Higuchi, J. C. Monge, N. Lee, S. W. Law and H. B. Brewer Jr, Apo B is encoded by a single copy gene in the human genome, Biochem. and Biophys. Res. Comm. 144:1332 (1987).

25. L. M. Powell, S. C. Wallis, R. J. Pease, Y. H. Edwards, T. J. Knott, J. Scott, A nowel form of tissue specific RNA processing produces apolipoprotein B 48 in intestine, Cell 50:831 (1987).

26. S. G. Young, S. J. Bertics, L. K. Curtiss, J. L. Witztum, Characterization of an abnormal species of apolipoprotein B, apolipoprotein B_{37}, associated with familial hypobetalipoproteinemia, J. Clin. Invest. 79:1831 (1987).

27. S. G. Young, S. J. Bertics, L. K. Curtiss, J. L. Witztum, Genetic analysis of a kindred with familial hypobetalipoproteinemia. Evidence for two separate gene defects : one associated with an abnormal apolipoprotein B species, apolipoprotein B_{37} ; and a second associated with low plasma concentrations of apolipoprotein B_{100}, J. Clin. Invest. 79:1842 (1987).

28. T. J. Knott, S. C. Rall, T. L. Innerarity, Human apolipoprotein B : structure of carboxyl-terminal domains, sites of gene expression, and chromosomal localization, Science 230:37 (1985).

29. T. J. Knott, R. J. Powell, S. C. Wallis, S. C. Rall, Complete protein sequence and identification of structural domains of human apolipoprotein B$_{100}$, Nature 323:734 (1986).

30. D. Hui, T. L. Innerarity, R. W. Milne, Y. L. Marcel, R. W. Mahley, Binding of chylomicron remnants and β-VLDL to hepatic and extra-hepatic lipoprotein receptors : a process independent of apolipoprotein B$_{48}$, J. Biol. Chem. 259:15060 (1984).

31. S. G. Young, F. P. Peralta, B. W. Dubois, L. K. Curtiss, J. K. Boyles, J. L. Witztum, Lipoprotein B$_{37}$, a naturally occuring lipoprotein containing the amino-terminal portion of apolipoprotein B$_{100}$, does not bind to the apolipoprotein B,E (low density lipoprotein) receptor, J. Biol. Chem. 262:16604 (1987).

32. S. G. Young, S. J. Bertics, T. M. Scott, B. W. Dubois, W. F. Beltz, L. K. Curtiss and J. L. Witztum, Apolipoprotein B allotypes MB19$_1$ and MB19$_2$ in subjects with coronary artery disease and hyper-cholesterolemia, Arteriosclerosis 7:61 (1987).

33. P. Duriez, N. Vu Dac, M. Koffigan, P. Puchois, C. Demarquilly, C. Fievet, P. Fievet, I. Luyeye, J. M. Bard, J. L. Fourrier, N. Slimane, J. M. Lablanche, M. Bertrand and J. C. Fruchart, Detection of human apolipoprotein B polymorphic species with one monoclonal antibody (BIP 45) against low density lipoprotein. Influence of this polymorphism on lipids levels and coronary artery stenosis, Atherosclerosis 66:153 (1987).

34. A. Barkia, C. Martin, P. Puchois, J. C. Gesquière, C. Cachera, A. Tartar and J. C. Fruchart, Enzyme-liked immunosorbent assay for human pro-apolipoprotein A-I using specific antibodies against synthetic peptide, J. Lipid Res. 29:77 (1988).